An Illustrated Book of Begonias

秋海棠属植物形态解剖图鉴

李凌飞　　杨蕾蕾　主编

Editors-in-chief　　Lingfei Li　　Leilei Yang

中国农业出版社
北　京

简　介
Introduction

　　秋海棠属植物是全球维管植物数量排名前五的大属之一，原生种超过2 100种，培育的品种更是数以万计。对于一个物种多样性如此丰富的大属，识别与鉴定并非一件易事。植物的鉴定往往涉及精细的形态结构描述，但鉴于秋海棠属植物种类繁多，早年发表的大部分种类更是缺乏详细的形态图片，不管对专业人士还是爱好者进一步探究都带来了诸多的不便。《秋海棠属植物形态解剖图鉴》是目前国际上唯一一本有关秋海棠属植物形态解剖的专著，全书首次展示了全球95种秋海棠属原生种类及32个栽培品种，每个种类包含了植株的整体形态、茎、叶、花等各个器官的细节。为了保证科学性，全书根据秋海棠属植物的全球分布情况，并基于目前使用最广泛的马克·休斯（Mark Hughes）最新分类系统（2024）对原生种进行了分组编排。本书不仅适用于植物学、园艺学等相关专业研究人员及大中专院校师生，普通读者也可以通过本书了解秋海棠属植物识别的基本知识，为鉴赏、保护、栽培和开发利用秋海棠属植物提供更加全面的基础。

Begonia is one of the world's top five largest genera of vascular plants, with more than 2,100 species and countless cultivars. Identification and classification can be challenging, given the rich diversity within this genus. Plant identification often involves detailed descriptions of morphological structures. Professionals and enthusiasts have faced difficulties in further exploration due to the vast number of Begonias and the lack of clear, detailed morphological images in early publications. This publication An Illustrated Book of Begonias is currently the only monograph worldwide dedicated to the morphological anatomy of the begonia family. The book showcases, for the first time, 95 species and 32 cultivars of Begonia. Each entry includes details of the overall plant morphology, stems/rhizomes, leaves, flowers, and other organs. To ensure scientific accuracy, the book categorizes species of Begonia according to the global distribution and the latest widely adopted classification system by Mark Hughes (2024). This book is suitable for researchers and students in botany, horticulture, and related fields and for the general public. Readers can gain basic knowledge of Begonia identification, providing a comprehensive foundation for appreciating, conserving, growing, and developing new hybrids of these fascinating plants.

An Illustrated Book of Begonias

Committee

CONTENTS 目录

1

秋海棠属植物形态解剖图鉴 An Illustrated Book of Begonias

秋海棠品种 Cultivars

竹节类 Cane-like

大王秋海棠栽培群 Rex-cultorum Group

根茎类 Rhizomatous

概　述
Summary

秋海棠属（*Begonia*）是葫芦目（Cucurbitales）秋海棠科（Begoniaceae）下的一个属。按最新的分类系统，秋海棠科仅包括2个属，即秋海棠属和夏威夷秋海棠属（*Hillebrandia*）。其中，夏威夷秋海棠属是分布在夏威夷群岛的一个单种属。与之形成鲜明对比，秋海棠属则成为全球维管植物总数排名前五的大属之一，数量十分庞大，目前全球已知的原生种超过2 100种。因此，秋海棠属几乎代表了整个秋海棠科。

The genus *Begonia* belongs to the family Begoniaceae in the order Cucurbitales. According to the latest classification system, Begoniaceae comprises only two genera: *Begonia* and *Hillebrandia* (endemic to the Hawaiian Islands). Notably, *Hillebrandia* consists of a single species in the Hawaiian archipelago. In sharp contrast, *Begonia* ranks among the top five largest genera of vascular plants, with an extensive diversity. Currently, over 2,100 species are recognized within the genus *Begonia*. Consequently, *Begonia* essentially represents the entirety of the family Begoniaceae.

秋海棠属植物主要分布在泛热带地区，在非洲、亚洲及中南美洲的热带与亚热带区域均有它们的身影。非洲是秋海棠属的起源中心，随后通过平行辐射演化，分别向美洲和亚洲扩散。作为起源中心的非洲，现存约160种，而作为多样化演化中心的美洲和亚洲分别超过800种和1 000种。中国目前记录超过277个分类群（含253个种、7个天然杂种、3个亚种和14个变种），是世界上秋海棠属植物分布最多的国家之一。

Begonias are primarily distributed across the tropical and subtropical regions of Africa, Asia, and Central and South America. Africa is considered the center of origin for *Begonia*, which then underwent parallel radiations, spreading to the Americas and Asia. Africa, as the center of origin, currently accounts for about 160 species, while the Americas and Asia, as centers of diversification, host more than 800 species and over 1,000 species, respectively. In China, more than 277 taxa are currently recorded (including 253 species, seven natural hybrids, three subspecies, and 14 varieties), making it one of the most species-rich countries with *Begonia* worldwide.

秋海棠英文名字"*Begonia*"的由来，最早可以追溯到17世纪末。法国传教士、植物学家查尔斯·普吕米耶（Charles Plumier）在安的列斯群岛发现了6个新种，并用*Begonia*命名，以纪念西印度群岛的法属圣多明各行政长官米歇尔·贝贡（Michel Bégon）。1700年，约瑟夫·皮顿·德图尔福尔（Joseph Pitton de Tournefort）在其著作中首次使用了*Begonia*这一名称。1753年，卡尔·林奈（Carl Linnaeus）将之前的6个种合并成一个，命名为斜叶秋海棠（*Begonia obliqua*），并将其收录在《植物种志》（*Species Plantarum*）中，沿用至今。

The English name 'Begonia' can be traced back to the late 17th century. The French monk and botanist Charles Plumier discovered six new species in the Antilles (Caribbean Islands). He named them 'Begonia' in honor of Michel Bégon, the French governor of Santo Domingo in the West Indies. The name 'Begonia' was first used in a publication by Joseph Pitton de Tournefort in 1700. In 1753, Carl Linnaeus merged the previously identified six species into one, naming it *Begonia obliqua*. This name was included in his work, *Species Plantarum*, and has been used ever since.

在中国，秋海棠经常与海棠混淆。从古籍记载来看，秋海棠一词的由来，最早可以追溯到南宋时期，当时有不少诗词出现了"秋日海棠"，但大多并非现在我们熟知的秋海棠。直到明中期，秋海棠的名称才真正开始被明确记载和描述。如边贡的《秋海棠》和陈道复的《题秋海棠》等。不过，坊间最为盛传的还是关于一妇人因思念丈夫，化泪为草（秋海棠）的典故以及南宋大诗人陆游与唐琬的凄美爱情故事，也因此成为了中国爱情花之一。此外，清末民初时期的中国地图从形态上与秋海棠叶片有点相似，即有了"我国地形，如秋海棠叶。出渤海，如叶之茎；西至葱岭，如叶之尖；各省及藩属，合为全叶"的描述。这将国家地理与之紧密相连，赋予其更为丰富的文化内涵。

In China, the 'Begonia' (Qiū Hǎi Táng), which translates to 'Autumn (Qiū) malus (Hǎi Táng)' is often mistaken for malus (apple). According to historical records, the term 'Begonia' can be traced back to the Southern Song Dynasty. During that period, many poems and literary works mentioned 'Autumnal malus', but most were not the begonia we are familiar with today. It wasn't until the mid-Ming Dynasty that the name 'Begonia' began to be recorded and described. For example, Gong Bian wrote a poem titled *Begonia*, and Daofu Chen wrote an *Inscription on Begonia*. However, the most widely circulated story is about a woman who turned her tears into grass (a reference to *Begonia grandis*) out of longing for her husband, and the poignant love story between the great poet You Lu and Wan Tang. Consequently, the begonia became one of the Chinese symbolic flowers of love. In addition, the maps of China in the late Qing Dynasty and early Republic of China period were pictured and described in a way resembling a begonia leaf. It was often stated that 'the shape of our country is like that of a begonia leaf. Starting from the Bohai Sea, it resembles the base of the leaf; extending west to the Pamir high plateau (Cōnglǐng), it resembles the tip of the leaf; each province and territory combined form the entire leaf'. This connects begonia with our geography and history, giving all a deeper connection and a richer cultural connotation.

一、秋海棠属的起源与分布
Origin and Distribution of *Begonia*

由于缺乏化石证据，目前对秋海棠属植物的起源主要依赖分子系统学结果进行推断。一般认为，秋海棠属起源发生在2 000万—6 500万年古新世（Paleogene）—中新世（Miocene）之间，且发生在中新世（mid-Miocene）—渐新世（Oligocene）可能性最大。大量研究表明，秋海棠属起源于非洲，但非洲现存的大部分种类则形成于更新世（Pleistocene）；而泛热带分布的其他秋海棠属植物可能是由于古代超大陆冈瓦纳的分裂后，通过洲际扩散而出现的。

Currently, there is still some controversy surrounding the origin time of the *Begonia*. Due to the lack of reliable fossil evidence, the inference is primarily based on phylogenies. It is generally believed that the *Begonia* originated between 20 and 65 million years ago (from the Paleogene to the Miocene), with the mid-Miocene to Oligocene being the most likely period. Numerous studies indicate that while Africa is considered the center of origin for *Begonia*, most of the extant species likely evolved during the Pleistocene epoch. The extant pan-tropical distribution of *Begonia* might have been facilitated by intercontinental dispersal following the breakup of ancient supercontinent Gondwana.

中新世是一个温暖的时期，是很多热带植物向亚洲扩张的重要历史时期。分子证据估算，亚洲秋海棠的起源时间在 1 500 万—1 800 万年之前，这与中新世气候适宜期相一致。秋海棠属在渐新世已经在非洲开始多样化。一种观点认为，当前秋海棠属植物在中国及亚洲东南部多样性热点的形成是非洲祖先通过喜马拉雅山脉向东扩散和迁移的结果。喜马拉雅山脉是亚洲板块和印度板块于 3 500 万年前相撞后形成的，而秋海棠属祖先到达的时间推测在 1 500 万年前，即与喜马拉雅山最高海拔出现的时间基本一致。因此，根据气候的特征推测，在印度次大陆北部的山地生境可能存在很大一块区域为秋海棠向东迁移提供了纽带，使中南半岛和马来群岛成为秋海棠属植物多样性热点地区。

The mid-Miocene was a warm period, making a significant historical period for expanding the range of many tropical plants into Asia. Molecular evidence suggests that the origin of Asian Begonias dates back more than 15-18 million years ago, which aligns with the favorable climatic conditions of the mid-Miocene. *Begonia* had already started diversifying in Africa during the Oligocene. One view is that the formation of the current diversity hotspots of *Begonia* in China and Southeast Asia resulted from the migration and expansion of African ancestors through the Himalayas to the east. The Himalayan mountain range formed when the Asian and Indian tectonic plates collided 35 million years ago. The estimated time for the ancestors of *Begonia* to have arrived in the area is about 15 million years, which coincides with the emergence of the highest elevations in the Himalayas. Therefore, based on climatic characteristics, it is speculated that a significant area of the montane habitats at the north of the Indian continent might have acted as a bridge for *Begonia* migration toward the east, resulting in Indo-China and Malay Archipelago becoming diversity hotspots for *Begonia*.

秋海棠属植物向美洲扩散的时间可能与亚洲相近。在美洲，秋海棠属植物从墨西哥南部一直分布到阿根廷的最北端。最新的系统基因组学研究表明，非洲秋海棠属祖先向美洲区域扩散至少存在两次独立的事件，随后通过广泛的杂交和基因渐渗，促使这一区域的物种多样性演化。

　　The timing of the expansion into the Neotropical region of the Americas, is likely similar to that of Asia. In the Americas, Begonias are distributed from Mexico south to the northernmost tip of Argentina. The latest genomic studies suggest that the African ancestors of *Begonia* dispersed into the Neotropical region at least two separate times. Subsequent extensive hybridization and gene introgression contributed to the evolution of species diversity in this region.

二、秋海棠属分类系统
Taxonomy of *Begonia*

　　秋海棠属植物大部分生境单一，特有性极高。对秋海棠属的分类，早期主要依赖于植株类型、叶片形态、花器官形态、子房室数、胎座类型、生活习性等形态学数据。自19世纪以来，有很多学者对其进行了处理。1841年，查尔斯·高迪豪德-比普利（Charles Gaudichaud-Beaupre）建立了肉果属（*Mezierea*）。1846年，约翰·林德利（John Lindley）根据每室2个胎座和4个花被片的特征，又分别建立了东亚秋海棠属（*Diploclinium*）和真瓣属（*Eupetalum*）。1855年，约翰·克洛奇（Johann Klotzch）根据雌花的柱头是否宿存这一特征，将秋海棠科分为两个族群，其中一个是宿存族（Stephanocarpeae），另一个为脱落族（Gymnocarpeae），并进一步根据花的形态特征将其划分为41个属，涵盖194个种，不过该处理方式并未得到大多数学者的认可。

Most of the habitats of Begonias are relatively singular, displaying high endemism. The classification of the *Begonia* was historically based on morphological data including plant types, leaf shapes, flower organ morphology, the number of ovary locules, placenta types, and habits. Since the 19th century, numerous scholars have attempted to classify it. In 1841, Charles Gaudichaud-Beaupre established the genus *Mezierea* based on specific characteristics. In 1846, John Lindley established the genera *Diploclinium* and *Eupetalum* based on features like two locules per ovary and four tetals. In 1855, Johann Klotzch divided the Begoniaceae family into Stephanocarpeae and Gymnocarpeae, based on whether the styles of the female flowers persisted. He divided them into 41 genera and 194 species based on flower morphology. However, these classification attempts were not widely accepted.

1859年，阿方斯·德康多尔（Alphonse de Candolle）在约翰·克洛奇系统的基础上对其进行了修订，最后将秋海棠科分成了肉果属、钩果属（Casparya）、秋海棠属3个属。其中，肉果属，3个种；钩果属，23个种；秋海棠属，323个种。1864年，他再次对秋海棠属进行研究，进一步将秋海棠属划分为61个组，其中有34个组与克洛奇系统对应，这些组大多数都狭域分布，限于非洲、亚洲和新热带地区的其中一个主要地理区域。这一处理方式随后被大多数分类学家采纳。

In 1859, Alphonse de Candolle revised the classification of the Begoniaceae family based on Johann Klotzch's system. He ultimately divided the family into three genera: *Mezierea*, *Casparya* (now included in the subgenus *Baryandra*), and *Begonia*. Among them are the genus *Mezierea* with three species, *Casparya* with 23 species, and *Begonia* with 323 species. In 1864, while studying the *Begonia*, de Candolle further divided the taxon into 61 sections, with 34 corresponding to

Klotzsch's system. Most of these sections were narrowly distributed, confined to one major geographic region such as Africa, Asia, or the Neotropics. Most taxonomists subsequently adopted this treatment.

1894年，奥托·瓦尔堡（Otto Warburg）在阿道夫·恩格勒（Adolf Engler）的《植物自然分科志》（*Die Natürlichen Pflanzenfamilien*）一书中将秋海棠科处理为4个属：即夏威夷秋海棠属、秋海棠属、合被秋海棠属（*Symbegonia*）和小秋海棠属（*Begoniella*），并首次根据分布地域将秋海棠属划分为12个非洲组、15个亚洲组和31个美洲组。

In 1894, Otto Warburg, in Adolf Engler's *Die Natürlichen Pflanzenfamilien*, reclassified the Begoniaceae family into four genera: *Hillebrandia*, *Begonia*, *Symbegonia*, and *Begoniella*. He also categorized *Begonia* into 12 African sections, 15 Asian sections, and 31 American sections based on their distribution.

1925年，埃德加·伊姆舍尔（Edgar Irmscher）在奥托·瓦尔堡的分类基础上进一步修订，扩展到了5个属，包括夏威夷秋海棠属（1种）、秋海棠属（含60个组，750种）、少蕊秋海棠属（*Semibegoniella*，2种，分布于厄瓜多尔）、小秋海棠属（3种，分布于哥伦比亚）和合被秋海棠属（10种，分布于新几内亚）。秋海棠属被划分为12个非洲组、16个亚洲组和32个美洲组。

In 1925, Edgar Irmscher further revised the classification based on Otto Warburg's work. He expanded to five genera, including *Hillebrandia* (1 species), *Begonia* (with 60 sections and 750 species), *Semibegoniella* (2 species, distributed in Ecuador), *Begoniella* (3 species, distributed in Colombia), and *Symbegonia* (10 species, distributed in New Guinea). *Begonia* was divided into 12 African sections, 16 Asian sections, and 32 American sections.

1998年，扬·多伦博斯（Jan Doorenbos）等人在《秋海棠分组》（*The Sections of Genus Begonia*）一文中，通过使用62个形态特征的数值分类法，将1 399个秋海棠属植物划分为63个组，这为后续的分类框架形成奠定了坚实的基础。

In 1998, Jan Doorenbos and his colleagues published a paper titled *The Sections of Genus Begonia*. This work used a numerical classification method based on 62 morphological characteristics to divide 1,399 Begonias into 63 sections. This study laid a foundation for subsequent developments in the classification framework.

伴随着分子系统学的广泛应用，被子植物系统发育研究组（Angiosperm Phylogeny Group，APG）系统逐步成为主流。2018年，彼得·穆恩莱特（Peter Moonlight）等人使用3个叶绿体片段（*ndhA intron, ndhF-rpl32 spacer, rpl32-trnL spacer*）构建了一个包括574个秋海棠科植物的系统发育树，将秋海棠属分为70个组，包括新增的5个组：星毛组（*Astrothrix*）、四季秋海棠组（*Ephemera*）、根茎单座组（*Jackia*）、匍茎组（*Kollmannia*）和星蕊组（*Stellandrae*），同时恢复了4个组：即球根秋海棠组（*Australes*）、非洲无翅组（*Exalabegonia*）、宽柱组（*Latistigma*）和美洲盾叶组（*Pereira*）。该分类处理也成了当前最广泛使用的系统。

The Angiosperm Phylogeny Group (APG) system gradually became mainstream with the widespread application of molecular phylogenetics. In 2018, Peter Moonlight *et al.* used three chloroplast DNA markers (*ndhA intron, ndhF-rpl32 spacer, rpl32-trnL spacer*) to construct a phylogenetic tree encompassing 574 species of the Begoniaceae. They divided the *Begonia* into 70 sections, adding five new sections: *Astrothrix, Ephemera, Jackia, Kollmannia,* and *Stellandrae*. They also reinstated four sections: *Australes, Exalabegonia, Latistigma,* and *Pereira*. This classification treatment has become the most widely used system to date.

2019年，税玉民等人选取了98个代表性的种类及7个外类群，利用叶绿体基因组序列进行了系统树构建，根据得到的结果，将秋海棠属分为14个亚属及48个组，包含新增的9个组：肿节组（*Biosiana*）、巨苞组（*Gigabracteata*）、海南秋海棠组（*Hainania*）、马拉巴尔秋海棠组（*Malabarae*）、鼠纹秋海棠组（*Murina*）、亚洲蕨叶组（*Pteridiformis*）、走茎组（*Stolonifera*）、管苞组（*Tubibracteolea*）、*Wuana*（密鳞组）；恢复了4个组：柱状秋海棠组（*Cylindribegonia*）、异型组（*Dysmorphia*）、簇毛组（*Flocciferae*）、印度秋海棠组（*Mitscherlichia*）。相比穆恩莱特（Moonlight）等人的系统中出现一些组为多系群，该系统所有的种间分类单元都是单系群，并且大多数分类单元可以通过地理起源、生态类型和物种形成模式来解释。

In 2019, Yu-Min Shui *et al*. selected 98 representative species and seven outgroups, and used chloroplast genome sequences to construct a systematic tree. Based on the results, they divided the *Begonia* into 14 subgenera, as well as 48 sections, including the addition of nine new sections: *Biosiana, Gigabracteata, Hainania, Malabarae, Murina, Pteridiformis, Stolonifera, Tubibracteolea, Wuana*. Additionally, four sections were reinstated: *Cylindribegonia, Dysmorphia, Flocciferae, Mitscherlichia*. In contrast to Moonlight *et al*.'s system, where some sections appeared to be polyphyletic, all inter-specific taxonomic units were monophyletic in this system. Their geographical origin, ecological type, and species speciation could explain most of these taxonomic units.

随着测序成本的降低及基因组测序的普及，为秋海棠属植物分类提供了新的发展机遇。2022年，李凌飞等人利用全基因组测序对覆盖37个组的全球78种代表性秋海棠属植物进行了全基因组测序，利用基因组数据，分别对单拷贝核基因和质体基因组进行系统树构建，发现核基因树与质体基因组系统树存在较大的差异，并通过基因流与渐渗分析（ABBA-BABA）及系统发育网络推断（Phylonet network）等证实了秋海棠属植物祖先存在广泛的杂交和基因渐渗事件，为解答秋海棠属的科学的分类提供了新的线索；同时提示我们需要更谨慎对待目前基于质体基因组/片段的分子系统学的结果。

With the decreasing cost of sequencing, new opportunities for the classification of *Begonia* have emerged through genome sequencing. In 2022, Lingfei Li *et al.* conducted whole-genome sequencing on 78 representative species of *Begonia*, covering 37 sections. They used genome data to construct phylogenetic trees based on single-copy nuclear genes and plastid genomes. They discovered significant differences between the nuclear and plastid genome trees. Through analyses such as ABBA-BABA tests and Phylonet network inference, they confirmed extensive hybridization and gene flow events in the ancestry of *Begonia*. This provides new insights for classification. Simultaneously, it suggests a need for a cautious interpretation of current molecular systematic results based on plastid genomes/fragments.

2022年，由爱丁堡皇家植物园的马克·休斯（Mark Hughes）牵头成立了秋海棠系统发育研究组（Begonia Phylogeny Group，BPG）。该研究组由国际上一群专注于研究秋海棠属的研究人员和科学家组成，主要目标是使用各种分子技术，如DNA测序和系统发育分析，研究秋海棠属内的系统发育关系，并基于可靠的系统发育数据建立一个稳定且可靠的亚属分类，来了解秋海棠物种的进化历史、地理分布和生态适应等。目前该研究组已经取得一定的研究进展。相信随着更广泛的取样，并结合形态学、生物地理学数据，在不久的将来，这个有趣且快速进化的植物大属分类将会最终得以解决，并为进一步的科学研究、引种保育及园艺开发应用奠定坚实的基础。

In 2022, the Begonia Phylogeny Group (BPG) was established, led by Mark Hughes from the Royal Botanic Garden Edinburgh. This group comprises a team of researchers and scientists from around the world dedicated to studying *Begonia*. Their main objective is to investigate phylogenetic relationships using molecular techniques such as DNA sequencing and phylogenetic analysis. They aim to establish a stable and reliable subgeneric classification based on robust phylogenetic data. This classification will provide insights into *Begonia* species' evolutionary history, geographic distribution, and ecological adaptations. As of this writing, the research group has made significant progress. With broader sampling and the integration of morphological and biogeographical data, it is believed that the classification of this diverse and rapidly evolving plant genus will be resolved soon. This will lay a foundation for further scientific research, conservation efforts, and horticultural developments.

秋海棠属最新分组表
The latest classification of sections in the genus *Begonia*

序号 No.	组别 Section	中文名 Chinese name	分布 Distribution	物种数量 Number of species
1	Sect. *Alicida* C. B.Clarke	裂翅组	亚洲 Asia	6
2	Sect. *Apterobegonia* Warb.	亚洲无翅组	亚洲 Asia	2
3	Sect. *Apteron* C. DC.	美洲无翅组	南美洲 South America	4
4	Sect. *Astrothrix* Moonlight	星毛组	南美洲 South America	6
5	Sect. *Augustia* (Klotzsch) A. DC.	无苞组	非洲 Africa	12
6	Sect. *Australes* L.B.Sm. & B. G. Schub.	球根秋海棠组	南美洲 South America	14
7	Sect. *Baccabegonia* Reitsma	浆果组	非洲 Africa	2
8	Sect. *Barya* (Klotzsch) A. DC.	合蕊组	南美洲 South America	1
9	Sect. *Baryandra* A. DC.	菲律宾秋海棠组	亚洲、大洋洲 Asia, Oceania	95
10	Sect. *Begonia* L.	秋海棠组	北美洲、南美洲 North & South Americas	46
11	Sect. *Boisiana* Y. M. Shui, W. H. Chen & W. K. Dong	肿节组	亚洲 Asia	4
12	Sect. *Bracteibegonia* A. DC.	苞片组	亚洲 Asia	16
13	Sect. *Casparya* (Klotzsch) Warb.	钩果组	北美洲、南美洲 North & South Americas	52

（续）

序号 No.	组别 Section	中文名 Chinese name	分布 Distribution	物种数量 Number of species
14	Sect. *Chasmophila* J. J. de Wilde & Plana	岩生组	非洲 Africa	1
15	Sect. *Coelocentrum* Irmsch.	侧膜组	亚洲 Asia	92
16	Sect. *Cristasemen* J. J. de Wilde	冠籽组	非洲 Africa	1
17	Sect. *Cyathocnemis* (Klotzsch) A. DC.	合苞组	南美洲 South America	24
18	Sect. *Diploclinium* (Lindl.) A. DC.	东亚秋海棠组	亚洲、大洋洲 Asia, Oceania	100
19	Sect. *Donaldia* (Klotzsch) A. DC.	美洲双瓣组	北美洲、南美洲 North & South Americas	2
20	Sect. *Doratometra* (Klotzsch) A. DC.	矛果组	北美洲、南美洲 North & South Americas	9
21	Sect. *Ephemera* Moonlight	四季秋海棠组	北美洲、南美洲 North & South Americas	12
22	Sect. *Erminea* A. DC.	短茎组	非洲 Africa	20
23	Sect. *Eupetalum* (Lindl.) A. DC.	真瓣组	南美洲 South America	16
24	Sect. *Exalabegonia* Warb.	非洲无翅组	非洲 Africa	2
25	Sect. *Filicibegonia* A. DC.	非洲蕨叶组	非洲 Africa	8
26	Sect. *Flocciferae* N.Krishna & Pradeep	簇毛组	亚洲 Asia	4
27	Sect. *Gaerdtia* (Klotzsch) A. DC.	竹节秋海棠组	北美洲、南美洲 North & South Americas	8
28	Sect. *Gireoudia* (Klotzsch) A. DC.	根茎秋海棠组	北美洲、南美洲 North & South Americas	113
29	Sect. *Gobenia* A. DC.	藤蔓组	南美洲 South America	11
30	Sect. *Haagea* (Klotzsch) A. DC.	亚洲双瓣组	亚洲 Asia	2
31	Sect. *Hainania* Y.M.Shui & W. H. Chen	海南秋海棠组	亚洲 Asia	1
32	Sect. *Hydristyles* A. DC.	多柱组	南美洲 South America	10
33	Sect. *Jackia* M. Hughes	根茎单座组	亚洲 Asia	69
34	Sect. *Knesebeckia* (Klotzsch) A. DC.	立茎组	北美洲、南美洲 North & South Americas	49
35	Sect. *Kollmannia* Moonlight	匍茎组	南美洲 South America	2
36	Sect. *Latistigma* A. DC.	宽柱组	南美洲 South America	5
37	Sect. *Lauchea* (Klotzsch) A. DC.	肉翅组	亚洲 Asia	7
38	Sect. *Lepsia* (Klotzsch) A. DC.	丛茎组	北美洲、南美洲 North & South Americas	9
39	Sect. *Loasibegonia* A. DC.	刺莲花秋海棠组	非洲 Africa	21

（续）

序号 No.	组别 Section	中文名 Chinese name	分布 Distribution	物种数量 Number of species
40	Sect. *Malabarae* Y.M.Shui & W.H.Chen	马拉巴尔秋海棠组	亚洲 Asia	1
41	Sect. *Mezierea* (Gaudich.) Warb.	肉果组	非洲 Africa	4
42	Sect. *Microtuberosa* Moonlight & Tebbitt	小块茎组	南美洲 South America	1
43	Sect. *Monophyllon* A. DC.	独叶组	亚洲 Asia	4
44	Sect. *Muscibegonia* A. DC.	藓状秋海棠组	非洲 Africa	2
45	Sect. *Nerviplacentaria* A. DC.	纤座组	非洲 Africa	10
46	Sect. *Oligandrae* M.Hughes & W. N. Takeuchi	少蕊组	亚洲、大洋洲 Asia, Oceania	6
47	Sect. *Parietoplacentalia* Ziesenh.	美洲侧膜组	北美洲、南美洲 North & South Americas	3
48	Sect. *Parvibegonia* A. DC.	小秋海棠组	亚洲 Asia	27
49	Sect. *Peltaugustia* (Warb.) F. A. Barkley	索科特拉秋海棠组	非洲 Africa	2
50	Sect. *Pereira* Brade	美洲盾叶组	南美洲 South America	4
51	Sect. *Petermannia* (Klotzsch) A. DC.	等翅组	亚洲、大洋洲 Asia, Oceania	493
52	Sect. *Pilderia* (Klotzsch) A. DC.	双苞组	北美洲、南美洲 North & South Americas	6
53	Sect. *Platycentrum* (Klotzsch) A. DC.	扁果组	亚洲 Asia	237
54	Sect. *Pritzelia* (Klotzsch) A. DC.	巴西秋海棠组	北美洲、南美洲 North & South Americas	167
55	Sect. *Putzeysia* (Klotzsch) A. DC.)	珠芽组	亚洲 Asia	1
56	Sect. *Quadrilobaria* A. DC.	四裂组	非洲 Africa	21
57	Sect. *Quadriperigonia* Ziesenh.	美洲四被组	北美洲、南美洲 North & South Americas	20
58	Sect. *Reichenheimia* (Klotzsch) A. DC.	单座组	亚洲 Asia	26
59	Sect. *Ridleyella* Irmsch.	亚洲盾叶组	亚洲 Asia	7
60	Sect. *Rossmannia* A. DC.	长翅组	南美洲 South America	1
61	Sect. *Rostrobegonia* Warb.	喙翅组	非洲 Africa	9
62	Sect. *Ruizopavonia* A. DC.	羽脉组	北美洲、南美洲 North & South Americas	32
63	Sect. *Scutobegonia* Warb.	非洲盾叶组	非洲 Africa	10
64	Sect. *Sexalaria* A. DC.	六翅组	非洲 Africa	1

（续）

序号 No.	组别 Section	中文名 Chinese name	分布 Distribution	物种数量 Number of species
65	Sect. *Solananthera* A. DC.	茄蕊组	南美洲 South America	3
66	Sect. *Squamibegonia* Warb.	鳞屑组	非洲 Africa	3
67	Sect. *Stellandrae* Moonlight	星蕊组	南美洲 South America	1
68	Sect. *Stolonifera* Y.M.Shui & W.H.Chen	走茎组	亚洲 Asia	10
69	Sect. *Symbegonia* (Warb.) L.L.Forrest & Hollingsw.	合被组	亚洲、大洋洲 Asia, Oceania	18
70	Sect. *Tetrachia* Brade	四室组	南美洲 South America	19
71	Sect. *Tetraphila* A. DC.	非洲四被组	非洲、亚洲 Africa, Asia	31
72	Sect. *Trachelocarpus* (Müll.Berol.) A. DC.	颈果组	南美洲 South America	4
73	Sect. *Urniformia* Ziesenh.	瓮果组	南美洲 South America	1
74	Sect. *Wageneria* (Klotzsch) A. DC.	攀缘组	北美洲、南美洲 North & South Americas	5
75	Sect. *Ignota*	未划分组	非洲、南美洲、亚洲、大洋洲 Africa, South America, Asia, Oceania	79

注：分组依据参照马克·休斯（Mark Hughes）创建的秋海棠资源中心（https://padme.rbge.org.uk/Begonia/home），数据截至2024年4月。

Note: The classification of sections is based on Begonia Resource Centre (https://padme.rbge.org.uk/Begonia/home) created by Mark Hughes, with data updated to April 2024.

三、秋海棠引种栽培与育种历史
Introduction, Cultivation, and Breeding History of Begonias

中国具有丰富的野生秋海棠属植物资源，以秋海棠（*Begonia grandis*）为代表的秋海棠属植物在中国有超过千年的栽培历史，但由于一直没有引起足够的重视，直到20世纪中后期才陆续有植物园开展引种收集工作。相比之下，欧美国家尽管缺少秋海棠属植物资源，但引种收集工作却走在前面。1777年，英国的威廉·布朗（William Brown）将牙买加的小秋海棠（*Begonia minor*）引种到邱园，开启了秋海棠属植物的栽培历史。

China possesses rich wild resources of *Begonia*, with *Begonia grandis* being a representative species that has been cultivated for over a thousand years. However, adequate attention has not been paid to the systematic introduction and collection of Begonias in China. In contrast, despite the lack of native *Begonia* resources there, European and American countries have taken the lead in the introduction and collection of these resources. In 1777, William Brown from Britain introduced the *Begonia minor* from Jamaica to Kew Gardens, marking the beginning of the cultivation history of *Begonia*.

由于早期受到长距离运输条件的限制，直到1835年，英国人纳撒尼尔·巴格肖·沃德（Nathaniel Bagshaw Ward）发明了便携式的玻璃温箱——沃德箱（Wardian case），为植物的长距离运输开辟了新途径，随后大量来自中南美洲的秋海棠属植物被引入欧洲，如粗茎秋海棠（*Begonia crassicaulis*）、华丽秋海棠（*Begonia luxurians*）、贝叶秋海棠（*Begonia conchifolia*）、帝王秋海棠（*Begonia imperialis*）等。到1847年，欧洲栽培的秋海棠种类已经达到了约80种。从19世纪50年代开始，来自中国、印度等亚洲的秋海棠被逐渐引入欧洲，并作为亲本，逐渐用于栽培品种的培育。

Due to the limitations of long-distance transportation in the early days, it was not until 1835 that Nathaniel Bagshaw Ward, an Englishman, invented the portable glass terrarium known as the "Wardian case". This innovation allowed the successful shipment of live specimens to faraway places. Subsequently, many Begonias from Central and South America were introduced to Europe, including *Begonia crassicaulis*, *Begonia luxurians*, *Begonia conchifolia*, and *Begonia imperialis*. By 1847, the number of Begonias cultivated in Europe had reached 80. Starting from the 1850s, Begonias from Asia, including China and India, were gradually introduced to Europe. These Asian species became parental sources for breeding cultivars.

　　1932年美国秋海棠协会成立，标志着秋海棠的引种进入了新的时代。欧美国家的引种变得更加频繁有序，涌现了大批秋海棠爱好者，墨西哥的睫毛秋海棠（*Begonia bowerae*）、巴西的奥尔森秋海棠（*Begonia olsoniae*）、马达加斯加的博格纳秋海棠（*Begonia bogneri*）、菲律宾的波利略（*Begonia polilloensis*）等大量来自全球的野生秋海棠资源被引入欧美。学会还通过资金资助的方式鼓励专业人士开展引种保育工作，促进了秋海棠野生资源的保护。

The establishment of the American Begonia Society (ABS) in 1932 marked the beginning of a new era for Begonias. The introduction and cultivation of Begonias in Western countries became more frequent and organized. A wave of Begonias enthusiasts emerged, when native species from around the world were introduced to Europe and the Americas. Species like *Begonia bowerae* from Mexico, *Begonia olsoniae* from Brazil, *Begonia bogneri* from Madagascar, and *Begonia polilloensis* from the Philippines were among the many introduced. The ABS also encouraged conservation efforts through financial support, incentivizing professionals to cultivate and conserve Begonias. This approach aimed to contribute to the protection of Begonias in the wild.

　　进入21世纪，亚洲大量的野生秋海棠资源被发现，吸引了全球的秋海棠专业机构和爱好者。除了欧美传统的秋海棠保育机构，如美国沃斯堡植物园、英国爱丁堡皇家植物园等，亚洲尤其是中国也涌现出一批优秀的保育机构，如台湾辜严倬云植物保种中心、中国科学院昆明植物研究所、上海辰山植物园、深圳市中国科学院仙湖植物园、厦门市园林植物园、中国科学院西双版纳热带植物园等，引种保育的秋海棠种类都在400种以上，大大促进了秋海棠属的科学研究和保育工作。与此同时，在中国也出现大量的"秋海棠迷"，他们利用各种途径从全世界各地获得秋海棠，并依靠网络交流平台进行交流分享。其中，由董文珂创办的爱棠（iBegonia）公众号为公众深入了解秋海棠属植物提供了专业科普平台，促进了秋海棠粉丝的交流。

Entering the 21st century, many wild Begonias were discovered in Asia, attracting the attention of Begonias enthusiasts and professional institutions worldwide. In addition to traditional Begonias conservation organizations in Europe and America (notably Fort Worth Botanic Garden in the USA and Royal Botanic Garden Edinburgh in the UK), Asia, especially China, has seen the emergence of outstanding conservation institutions. These include the Dr. Cecilia Koo Botanic Conservation Center (KBCC) in Taiwan, Kunming Institute of Botany, Chinese Academy of Sciences (CAS), Shanghai Chenshan Botanical Garden, Shenzhen Fairy Lake Botanical Garden Shenzhen & CAS, Xiamen Botanical Garden, and Xishuangbanna Tropical Botanical Garden, CAS. These institutions have collectively introduced and conserved over 400 Begonias, significantly advancing scientific research and conservation efforts for the genus. Simultaneously, a large number of Begonia enthusiasts have emerged in China. They acquire Begonias through various channels worldwide and communicate and share their passion through online platforms. Notably, the iBegonia WeChat public account founded by Wen-Ke Dong has provided a professional platform for people to understand Begonias better, promoting communication among enthusiasts and facilitating their engagement.

2020年10月，中国野生植物保护协会秋海棠专业委员会成立，标志着中国秋海棠属植物的研究、保育和开发应用逐步走上正轨。该专业委员会每年举办相关的学术年会，邀请不同研究领域的秋海棠人分享自己的成果，并开展科普宣传、新品种开发工作，促进了中国秋海棠资源保护及开发利用。

In October 2020, the Begonia Committee (BC) of the China Wild Plant Conservation Association (CWPCA) was established, marking a significant step forward for the research, conservation, and practical applications of Chinese *Begonia* species. This committee annually organizes academic conferences, inviting experts from various fields related to Begonias to share their research findings. Additionally, the committee engages in science communication, popularization efforts, and developing new *Begonia* cultivars. These initiatives have advanced the conservation and utilization of Chinese *Begonia* resources.

在引种收集的基础上，秋海棠属的育种工作也如火如荼地进行着。秋海棠属植物种间杂交亲和力高，且结实率高、种子量大，因此非常利于杂交选育获得新品种。据不完全统计，自1845年首个杂交品种红叶秋海棠（*Begonia* 'Erythrophylla'）问世以来，目前全世界已经育成了1万多个栽培品种。其中，大部分的品种通过杂交得到，少量通过物理或化学诱变、自然芽变或自然变异优株选育获得，并相继形成了大王秋海棠栽培群、四季秋海棠栽培群、球根秋海棠栽培群、丽格秋海棠栽培群等4个主要栽培群。作为世界著名的观赏花卉之一，秋海棠占国际盆栽花卉市场总产量的第4位，并呈现出不断增长的趋势。

Begonias exhibit high interspecific hybrid compatibility within the genus, high fruiting rates and large seed quantities. As a result, they are exceptionally conducive to breeding new cultivars. According to incomplete statistics, since the introduction of the first hybrid cultivar, *Begonia* 'Erythrophylla', in 1845, more than 10,000 cultivars have been developed worldwide. Most of these cultivars are achieved through hybridization, while a smaller number resulted from physical or chemical mutagenesis, natural bud variations, or selection of superior naturally occurring variants. These efforts have led to the formation of four major cultivation groups: Rex-cultorum, Semperflorens-cultorum, Tuberous, and Rieger. Being one of the world's renowned ornamental flowers, Begonias rank fourth in total production of potted flowers in the international market, showing a continuously growing trend.

目前，市面上的品种主要由欧美、澳大利亚、日本等国家和地区培育，除几个大的育种公司如荷兰的科比（Koppe）公司、美国的特拉诺瓦（Terra Nova Nurseries）公司，很大一部分是由爱好者或科研机构培育的。相比之下，中国的育种起步较晚。中国科学院昆明植物研究所是国内最早开展育种研究的单位，从1995年开始，目前已经育成并登记31个品种。随后，上海辰山植物园、厦门园林植物园、桂林植物园、云南省农业科学院花卉研究所也相继培育了部分新品种。近年来，随着社会关注度提高，国内部分园艺花卉公司，如虹越花卉股份有限公司、北京市花木有限公司以及大量的业务爱好者加入育种队伍，使秋海棠育种在中国逐渐开始火热。

Currently, the cultivars available in the market are primarily developed in countries such as Europe, the USA, Australia, and Japan. While a few major breeding companies like Koppe from the Netherlands and Terra Nova from the USA are involved, enthusiasts or research institutions conduct a significant portion of hybridization. In comparison, China's breeding efforts started relatively later. The Kunming Institute of Botany was among the first domestic institutions to conduct breeding research in 1995. They have successfully bred and registered 31 cultivars to date. Subsequently, Shanghai Chenshan Botanical Garden, Xiamen Botanical Garden, Guilin Botanical Garden, Floriculture Research Institute, and Yunnan Academy of Agricultural Sciences have also introduced new cultivars through their breeding efforts. With increasing societal attention in recent years, several Chinese horticultural and flower companies, such as Hongyue Flower Co., Ltd., Beijing Florascape Co., Ltd., and numerous enthusiasts, have joined the breeding efforts. This has led to a gradual surge in the popularity of *Begonia* breeding in China.

四、秋海棠园艺学分类
Horticultural Classification of Begonias

随着市面上秋海棠属植物的种类越来越多，为了便于交流，园艺学家和普通消费者通常不会像分类学家那样细分每一个种类，他们往往根据植株形态及生长特性，对其进行划分。目前国际上比较通用的分类体系是由美国秋海棠协会制定的，将秋海棠属原生种及品种按植株形态及亲缘关系分为9类：

With the increasing variety of Begonias available in the market, it appears that horticulturists and ordinary consumers usually do not categorize each species as extensively as taxonomists do. They often classify them based on plant morphology and growth characteristics. Currently, the widely accepted classification system, used internationally was established by the American Begonia Society. It divides species and cultivars of *Begonia* into nine categories:

竹节类 Cane-like Begonias

株型如竹子，最高可长至3 ~ 5 m。全球有80余种原生种及2 000个栽培品种，但很多品种来源信息不详，很难鉴定。竹节类品种抗性较强，露天或办公室都适合种植，因此受到公众的热捧。最古老的一个竹节类品种是1890年育成的名为*Begonia* 'Corallina de Lucerna' 品种，至今仍流行。

These Begonias exhibit a bamboo-like growth habit, with some cultivars capable of reaching 3 to 5 meters. Globally, there are approximately 80 species in the wild and over 2,000 cultivars within this category. However, many of these cultivars lack detailed source information, making their identification challenging. Cane-like Begonias are known for their robust habit and can be grown outdoors or in office settings, making them popular among the public. One of the oldest cane-like cultivars is *Begonia* 'Corallina de Lucerna', developed in 1890 and remains popular today.

根茎类 Rhizomatous Begonias

该类型大多数不具有地上茎，地上部分可见的营养器官主要是叶柄和叶片，部分种类具有直立茎（Upright-jointed stem）。该类型叶形和叶色变化丰富，叶片大小从迷你型到巨型都有，有些种类叶片直径甚至可以达到近1 m。根茎类的品种也非常多，其中最有名的一个品种是红叶秋海棠（*Begonia* 'Erythrophylla'），该品种于1845年培育，是目前已知最古老的品种，被称为祖母秋海棠，至今仍在全世界广泛栽种。

Most species in this category lack aerial stems, and the visible vegetative organs are primarily the petioles and leaves. Some cultivars have upright-jointed stems. The foliage of Rhizomatous Begonias displays a wide range of shapes and colors. Leaf sizes can vary from miniature to giant, with some species having leaves that can reach a diameter of nearly 1 meter. There is a rich diversity of cultivars within the Rhizomatous Begonias. One of the most well-known cultivars is *Begonia* 'Erythrophylla', cultivated in 1845 and recognized as one of the oldest known *Begonia* cultivars. Referred to as the Grandmother of Begonia's, it continues to be widely cultivated worldwide.

丛生类 Shrub-like Begonias

该类型地上茎明显，通常呈灌木状，也有匍匐状的，与竹节类秋海棠相比，只是茎的形态并不像竹子，且大多数种类开花的频率和花量不能与竹节秋海棠相媲美，花色以白色为主。该类型代表性种类有金线秋海棠（*Begonia listada*）、棕榈叶秋海棠（*Begonia luxurians*）、多叶秋海棠（*Begonia foliosa*）。

The above-ground stems are evident in this category and typically take on a shrub-like or sometimes prostrate growth form. In comparison to the Cane-like Begonias, the stem morphology is not bamboo-like. Moreover, most species in this category do not match the flowering frequency and quantity of the Cane-like Begonias, and their flowers are predominantly white. Examples of this type include *Begonia listada*, *Begonia luxurians,* and *Begonia foliosa*.

粗茎类 Thick-Stemmed Begonias

该类型具有非常粗壮的茎干，茎基生，基本不分枝。随着茎的生长，基部的叶片会逐渐脱落，通常保留顶部较新的叶子，如双瓣秋海棠（*Begonia dipetala*）、开普敦秋海棠（*Begonia dregei*）。

This category features exceptionally stout stems that are basal and generally unbranched. As the stems grow, the lower leaves gradually drop off, leaving mainly the newer leaves at the top. Examples of this type include *Begonia dipetala* and *Begonia dregei*.

蔓生类 Trailing-Scandent Begonias

该类型具有典型的匍匐茎，呈蔓藤状，可攀援。栽培时可以提供树干、建筑物等供其攀爬，或用悬挂的容器，让其垂吊生长。该类型不多，杂交品种较少，目前大约有35个原种和35个栽培品种。常见代表性种类为气根秋海棠（*Begonia radicans*）。

This category features characteristic trailing stems with a creeping or vining growth habit. When cultivated, they can climb and cling to supports such as tree trunks or buildings or be grown in hanging containers, displaying a pendulous growth form. This category is less common, with fewer hybrid cultivars. Currently, there are approximately 35 species and 35 cultivars of this type. One popular examples is *Begonia radicans*.

球根秋海棠栽培群 Tuberhybrida Group Begonias

该类型具有明显的球茎，喜欢冷凉的环境，在生长季结束（通常为冬季）或气候干燥时会休眠，在南部地区通常因为温度过高，不适合其生长。中华秋海棠正是因为能通过球茎休眠的方式来抵御寒冷的冬季，才能分布到最北端。该类型的杂交品种几乎完全以中南美洲高海拔的种类杂交而成，大多具有花大、量多且重瓣的特性。比较出名的品种有金正日秋海棠（*Begonia × tuberhybrida* 'Kimjongilhwa'）。

Prominent tubers characterize this category. These plants prefer cool temperatures. They typically go dormant at the end of the growing season (usually in winter) or during periods of dry climate. In southern regions, these begonias may not thrive due to high temperatures. The Chinese native species *Begonia grandis*, for instance, can withstand cold winters through its tuberous dormancy, allowing it to grow even in the northernmost areas. Hybrid cultivars in this category are primarily derived from species found in high-altitude regions of Central and South America. They often possess ample, abundant, and densely petaled flowers. Notable cultivars include *Begonia × tuberhybrida* 'Kimjongilhwa'.

大王秋海棠栽培群 Rex-cultorum Group Begonias

这是一类以原产于亚洲的大王秋海棠（*Begonia rex*）及其杂交子代为亲本，通过与其他类型的秋海棠杂交得到的一系列品种，目前全球共超过 4 000 种。它们大多数叶色丰富，呈现银色、红色、绿色和紫色等。根据亲本的不同，叶片的形态也极为多样，有简单的，有枫叶状的，甚至螺旋状的。常见的种类有蜗牛秋海棠（*Begonia* 'Escargot'）。

This category includes a diverse group of Begonias primarily derived from the native Asian species *Begonia rex* and its hybrids, obtained through crosses with other types of begonias. There are currently over 4,000 cultivars globally. These begonias are known for their rich leaf colors, which shades of silver, red, green, and purple. The leaf shapes vary greatly depending on the parentage, ranging from simple to maple-like and even spiral forms. One popular example is *Begonia* 'Escargot'.

四季秋海棠栽培群 Semperflorens-cultorum Group Begonias

该类型秋海棠实际上是许多种类经过数千代与其他栽培品种进行交叉杂交得到的。它是目前全球范围最常见的，也是应用最广泛的一类秋海棠。它们抗性极强，耐热、耐阴湿，甚至还耐一定程度的寒冷，并且在不需要太多管养的条件下可以常年开花，在整个园艺界都可以称得上优秀。

This type of *Begonia* is, in fact, the result of thousands of generations of crossbreeding between numerous species and other cultivars. Semperflorens-cultorum Group Begonias are currently the most common and widely used type globally. They exhibit remarkable resistance, enduring heat, shade, and even moderate cold. Furthermore, they can bloom throughout the year with minimal care requirements, making them highly regarded throughout the horticultural community.

不少人可能会误认为有一种叫做四季秋海棠（*Begonia semperflorens*）的种类。的确早在19世纪，曾有一个种被如此命名，后来发现它实际上是兜状秋海棠（*Begonia cucullata*），随后这个名称被分类学家所废弃。然而，园艺家们却沿用这个名称，尽管它是错误的。

Many people might mistakenly believe that a species called *Begonia semperflorens* exists. Indeed, as early as the 19th century, a species was named this way. It was later renamed *Begonia cucullata*. Subsequently, this name was discarded by taxonomists. However, horticulturists have continued to use this name, even though it is incorrect.

以下是部分被用于培育四季秋海棠的亲本：兜状秋海棠、近绒毛秋海棠（*Begonia subvillosa*）、毛秋海棠（*Begonia hirtella*）、近纤毛秋海棠（*Begonia subciliata*）、瓦氏秋海棠（*Begonia wallichiana*）、红花秋海棠（*Begonia rubriflora*）、多叶秋海棠（*Begonia foliosa*）、费氏秋海棠（*Begonia fischeri*）。

The following are some of the parents that have been used to create the Semperflorens Begonias: *Begonia cucullata*, *Begonia subvillosa*, *Begonia hirtella*, *Begonia subciliata*, *Begonia wallichiana*, *Begonia rubriflora*, *Begonia foliosa*, and *Begonia fischeri*.

丽格秋海棠栽培群 Rieger Group Begonias

与四季秋海棠栽培群类似，该类型秋海棠也是一个由多种不同的亲本反复杂交得到的一类品种。主要是球根秋海棠和来自索科特拉秋海棠（*Begonia socotrana*），以及少量的开普敦秋海棠（*Begonia dregei*）、四季秋海棠等的杂交品种。

The Rieger Group is a cultivar group similar to Semperflorens-cultorum Group. It is a group of cultivars obtained through repeated hybridization of various parent species. It mainly involves *Begonia × tuberhybrida* and includes contributions from *Begonia socotrana*, with some incorporation of *Begonia dregei*, *Begonia × semperflorens*, etc.

索科特拉秋海棠为冬季开花，19世纪80年代育种者将它与球根秋海棠栽培品种（*Begonia × tuberhybrida*）进行杂交，获得了早期的丽格秋海棠品种，它们在冬季开花。虽然栽培难度较高，但由于其花色丰富、且全年开花，因此很快成为全球最流行的品种之一。

The winter-flowering *Begonia socotrana* was crossed with *Begonia × tuberhybrida* in the 1880s by breeders, resulting in the early Rieger Group cultivars that bloom during the winter. Despite being more challenging to cultivate, these cultivars quickly became one of the most popular cultivars globally due to their rich flower colors and year-round blooming.

此外，部分国家如英国、日本、中国等也有一些分类系统，但总体上分类更加简化，仅按茎的类型进行划分，分为：根茎类（Rhizomatous Begonias）、球茎类（Tuberous Begonias）、直立茎类（或须根类，Erect-stemmed Begonias or Fibrous Begonia）三大类。

Additionally, some countries like the UK, Japan, and China have their simplified classification systems, mainly categorizing Begonias based on stem types into three major groups: Rhizomatous Begonias, Tuberous Begonias, and Erect-stemmed Begonias (also known as Fibrous Begonias).

竹节类
Cane-like

根茎类
Rhizomatous

丛生类
Shrub-like

粗茎类
Thick-Stemmed

蔓生类
Trailing-Scandent

球根秋海棠栽培群
Tuberhybrida Group

大王秋海棠栽培群
Rex-cultorum Group

四季秋海棠栽培群
Semperflorens-
cultorum Group

丽格秋海棠栽培群
Rieger Group

秋海棠园艺学分类（参照美国秋海棠协会）
Horticultural Classification of Begonias
(Refer to the American Begonia Society)

五、秋海棠属植物形态特征
Morphological Characteristics of Begonias

秋海棠属植物为多年生肉质草本，偶有一年生草本或亚灌木，不同类型的植株形态大小存在明显差异。全球最大的秋海棠是产自南美洲的小花秋海棠（*Begonia parviflora*），植株高达超过 8 m；亚洲最大的秋海棠是原产于中国西藏墨脱的巨型秋海棠（*Begonia giganticaulis*），株高达 3.6 m；全球最小的秋海棠是极小秋海棠（*Begonia elachista*），成年植株高度不超过 4 cm。叶片大小差别也极为显著，原产墨西哥南部的大叶秋海棠（*Begonia fusca*）叶形巨大，直径可达 60 cm，可以遮住我们的半个身体；而一些小型种类，如极小秋海棠、小叶秋海棠（*Begonia parvula*），棱果秋海棠（*Begonia prismatocarpa*），叶片仅有指甲盖大小。

Begonias are perennial succulent herbs, occasionally annual herbs, or subshrubs. Significant differences exist in the size and morphology of various species. The world's largest begonia is *Begonia parviflora* from South America, with a height of more than 8 meters; *Begonia giganticaulis*, from Motuo, Xizang, China, reaches a towering height of 3.6 meters, making it the tallest known *Begonia* in Asia, and the world's smallest is *Begonia elachista*, with a height of no more than 4 centimeters. Leaf size is also diversified among Begonias. *Begonia fusca*, native to southern Mexico, has a leaf with a diameter of up to 60 centimeters, capable of concealing half of a person's body. In contrast, some small species, such as *Begonia elachista*, *Begonia parvula*, and *Begonia prismatocarpa*, have leaves no larger than a fingernail.

除了植株形态多样性外，秋海棠属植物在叶片形态和叶片斑纹等方面也呈现出极高的多样性。秋海棠叶片通常偏斜不对称，多数为单叶，少数种类为复叶类型。叶片形态从披针形、线形、椭圆形、心形、盾形、匙形、螺旋状到掌状不等，几乎涵盖了被子植物所有的叶片形态；叶缘从全缘、锯齿状、圆齿、波状、皱波状、浅裂至深裂或深裂后再裂等，形态变化复杂。叶色有银色、绿色、蓝色、红色等，叶片斑纹、斑点色彩形式多样，部分种类会呈现金属光泽，丝绒质感等。

秋海棠属植物叶片形态和斑纹多样性
Diversity in Leaf Morphology and Patterns of Begonias

A. 光滑秋海棠 *Begonia psilophylla*　B. 开普敦秋海棠 *Begonia dregei*　C. 红芒秋海棠 *Begonia thelmae*　D. 方氏秋海棠 *Begonia fangii*　E. 匙叶秋海棠 *Begonia blancii*　F. 金线秋海棠 *Begonia listada*　G. 波利略秋海棠 *Begonia polilloensis*　H. 气根秋海棠 *Begonia radicans*　I. 多叶秋海棠 *Begonia foliosa*　J. 帝王秋海棠 *Begonia imperialis*　K. 柔浩秋海棠 *Begonia jiewhoei*　L. 小兄弟秋海棠 *Begonia* 'Little Brother Montgomery'　M. 斑叶竹节秋海棠 *Begonia maculata*　N. 柳伯斯秋海棠 *Begonia lubbersii*　O. 列梅秋海棠 *Begonia thiemei*　P. 蛛网脉秋海棠 *Begonia arachnoidea*　Q. 克娄巴特拉秋海棠 *Begonia* 'Cleopatra'　R. 艳后秋海棠 *Begonia cleopatrae*　S. 黑峰秋海棠 *Begonia ferox*　T. 水鸭脚 *Begonia formosana*　U. 蜗牛秋海棠 *Begonia* 'Escargot'　V. 宁巴四翅秋海棠 *Begonia quadrialata* subsp. *nimbaensis*　W. 桑寄生状秋海棠 *Begonia loranthoides*　X. 布隆泽秋海棠 *Begonia* 'Bronze King'　Y. 彩纹秋海棠 *Begonia variegata*

In addition to the diversity in morphology, Begonias exhibit a high degree of variation in leaf morphology and variegation. *Begonia* leaves are typically obliquely asymmetrical, mainly of the simple leaf type, with a few species having compound leaves. Leaf shapes range from lanceolate, linear, elliptical, heart-shaped, peltate, spoon-shaped, helical, to palmate, covering nearly all leaf shapes found in angiosperms. Leaf margins vary from entire, serrate, dentate, undulate, crisped, shallowly lobed, to deeply lobed or doubly lobed, showcasing a complex array of morphological changes. The leaf color can be silver, green, blue, red, and more, with diverse patterns of veins, spots, and colors. Some species even exhibit a metallic sheen and a velvety texture.

秋海棠属植物叶片表面纹饰也极为丰富，有具褶皱或皱纹的、具泡状隆起、疤状突起的、还有一些种类表面具小孔或带一些均匀的小斑点。大多数种类叶片表面具有毛状物，根据毛状物的不同，可以分为绒毛状、毛状和长毛状；很多种类还长有腺毛，根据软硬程度，分为刚毛、柔毛、绒毛等。

The leaf surfaces of Begonias are also incredibly diverse in their decorations. Some exhibit rugose, bullate, and pustulate, while others have foveolate, or with a number of small, uniform spots on the surface. The surfaces of most species' leaves bear various types of hairs, categorized as tomentose, pubescent, and long-hairy, based on the nature of the hairs. Many species also feature glandular hairs, which can be classified as stiff, soft, or velvety depending on their texture.

秋海棠属植物花均为单性，多数种类为雌雄同株，少数雌雄异株，二歧聚伞花序腋生或顶生。花色丰富，涵盖白色、粉色、红色、绿色、黄色和橙色等，不少种类还常常伴有混杂色彩。秋海棠属植物花被片数目在2～8个之间，常为4，两大两小，呈十字对称；同一种的雌雄花的花被片数量也会存在差异。

Begonias typically have unisexual flowers, with most species being monoecious (having both male and female flowers on the same plant), while a few are dioecious (male and female flowers on separate plants). The inflorescences are axillary or terminal, cymes dichasial. The flowers display various colors, including white, pink, red, green, yellow, and orange, often with intricate blends. The number of tepals in *Begonia* flowers ranges from 2 to 8, commonly 4, with a characteristic cruciform symmetry (four tepals arranged like a cross). There can be variations in the number of tepals between male and female flowers of the same species.

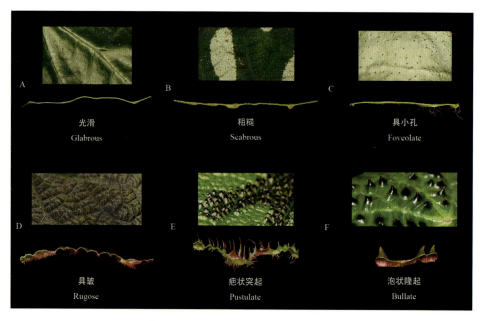

秋海棠属植物叶片表面形态多样性
Diverse Leaf Surface Morphology of Begonias

A. 光滑秋海棠 *Begonia psilophylla*　B. 涩叶秋海棠 *Begonia scabrifolia*　C. 靖西秋海棠 *Begonia jingxiensis*　D. 陆氏秋海棠 *Begonia lui*　E. 铁十字秋海棠 *Begonia masoniana*　F. 黑峰秋海棠 *Begonia ferox*

秋海棠属植物叶片表皮毛形态多样性
Diverse Leaf Surface Hairs in Begonias

A. 绿脉秋海棠 *Begonia chloroneura*　B. 榆叶秋海棠 *Begonia ulmifolia*　C. 红芒秋海棠 *Begonia thelmae*　D. 长纤秋海棠 *Begonia longiciliata*　E. 倬云秋海棠 *Begonia zhuoyuniae*　F. 镜毅秋海棠 *Begonia chingipengii*　G. 金线秋海棠 *Begonia listada*　H. 瓦氏秋海棠 *Begonia wallichiana*　I. 花叶秋海棠 *Begonia cathayana*　J. 广东秋海棠 *Begonia guangdongensis*　K. 汤姆森秋海棠 *Begonia thomsonii*　L. 蛛网脉秋海棠 *Begonia arachnoidea*　M. 帝王秋海棠 *Begonia imperialis*　N. 巴西变色秋海棠 *Begonia solimutata*　O. 宁明秋海棠 *Begonia ningmingensis*　P. 须苞秋海棠 *Begonia fimbribracteata*

　　秋海棠属植物的果实多为蒴果，少数为浆果状，其形状有椭圆形、长卵形、卵球形、三角形等；通常具有3个等长或不等长的翅，部分种类具4翅或无翅，翅的形态多样，有新月形、镰刀形，三角形、近矩形等。子房通常有1～4室不等，胎座类型为侧膜胎座或中轴胎座，因此，子房形态是秋海棠属植物分类鉴定的重要依据之一。

The fruits of Begonias are predominantly capsules, with a few being fleshy. They come in various shapes, including elliptical, elongated-ovoid, ovoid, and triangular. Typically, these capsules possess three equal or unequal wings, although some species may exhibit four wings or are wingless. Wings range from crescent-shaped and falcate to triangular or nearly rectangular. The ovary is usually 1 to 4 locular, with the placentae being parietal or axile. Therefore, the morphology of the ovary is a crucial characteristic for the identification and classification of Begonias.

秋海棠属植物花型和花色多样性
Diverse Flower Morphology and Colors in Begonias

A. 墨脱秋海棠 *Begonia hatacoa*   B. 汤姆森秋海棠 *Begonia thomsonii*   C. 大王秋海棠 *Begonia rex*   D. 迪特里希秋海棠 *Begonia dietrichiana*   E. 奥尔森秋海棠 *Begonia olsoniae*   F. 广东秋海棠 *Begonia guangdongensis*   G. 艳后秋海棠 *Begonia cleopatrae*   H. 红探戈秋海棠 *Begonia* 'Red Tango'   I. 西南秋海棠 *Begonia cavaleriei*   J. 方氏秋海棠 *Begonia fangii*   K. 川边秋海棠 *Begonia duclouxii*   L. 气根秋海棠 *Begonia radicans*   M. 珊瑚秋海棠 *Begonia* 'Pinafore'   N. 高茎秋海棠 *Begonia sphenantheroides*   O. 多花秋海棠 *Begonia sinofloribunda*   P. 刘演秋海棠 *Begonia liuyanii*   Q. 近革叶秋海棠 *Begonia subcoriacea*   R. 柳伯斯秋海棠 *Begonia lubbersii*   S. 鳞叶秋海棠 *Begonia scutifolia*   T. 钟扬秋海棠 *Begonia zhongyangiana*   U. 橙花侧膜秋海棠 *Begonia aurantiflora*

秋海棠属植物果实形态多样性
Diverse Fruit Morphology in Begonias

　　A. 柱果秋海棠 *Begonia cylindrica*　B. 桑寄生状秋海棠 *Begonia loranthoides*　C. 鳞叶秋海棠 *Begonia scutifolia*　D. 铺地秋海棠 *Begonia handelii* var. *prostrata*　E. 崇左秋海棠 *Begonia chongzuoensis*　F. 宁巴四翅秋海棠 *Begonia quadrialata* subsp. *nimbaensis*　G. 伯基尔秋海棠 *Begonia burkillii*　H. 假大新秋海棠 *Begonia pseudodaxinensis*　I. 榆叶秋海棠 *Begonia ulmifolia*　J. 软茎秋海棠 *Begonia mollicaulis*　K. 赤水秋海棠 *Begonia chishuiensis*　L. 长纤秋海棠 *Begonia longiciliata*　M. 纳土纳秋海棠 *Begonia natunaensis*　N. 巴西变色秋海棠 *Begonia solimutata*　O. 柳伯斯秋海棠 *Begonia lubbersii*　P. 兰屿秋海棠 *Begonia fenicis*　Q. 龙虎山秋海棠 *Begonia umbraculifolia*　R. 革叶秋海棠 *Begonia coriacea*　S. 绿脉秋海棠 *Begonia chloroneura*　T. 彩纹秋海棠 *Begonia variegata*　U. 钟扬秋海棠 *Begonia zhongyangiana*　V. 灯果秋海棠 *Begonia lanternaria*　W. 高茎秋海棠 *Begonia sphenantheroides*

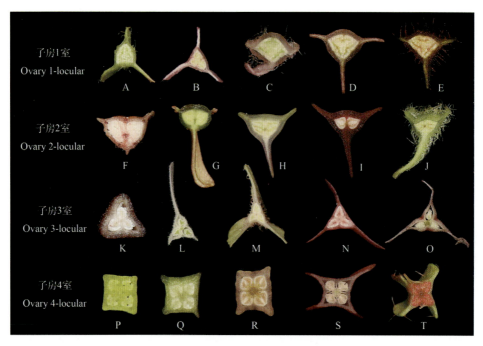

秋海棠属植物子房室数多样性

Diversity of Ovary Locules in Begonias

A. 橙花侧膜秋海棠 *Begonia aurantiflora*　B. 龙虎山秋海棠 *Begonia umbraculifolia* C. 丽叶秋海棠 *Begonia ningmingensis* var. *bella*　D. 弄岗秋海棠 *Begonia longgangensis* E. 刘演秋海棠 *Begonia liuyanii*　F. 山地秋海棠 *Begonia oreodoxa*　G. 长翅秋海棠 *Begonia longialata*　H. 九九峰秋海棠 *Begonia bouffordii*　I. 光滑秋海棠 *Begonia psilophylla*　J. 钟扬秋海棠 *Begonia zhongyangiana*　K. 癞叶秋海棠 *Begonia leprosa* L. 软茎秋海棠 *Begonia mollicaulis*　M. 刺萼秋海棠 *Begonia echinosepala*　N. 革叶秋海棠 *Begonia coriacea*　O. 龙胄秋海棠 *Begonia dracopelta*　P. 微籽秋海棠 *Begonia microsperma*　Q. 棱果秋海棠 *Begonia prismatocarpa*　R. 铺地秋海棠 *Begonia handelii* var. *prostrata*　S. 伯基尔秋海棠 *Begonia burkillii*　T. 宁巴四翅秋海棠 *Begonia quadrialata* subsp. *nimbaensis*

参考文献
References

Ardi, W, Campos-Dominguez, L, Chung, K F, et al., 2022. Resolving phylogenetic and taxonomic conflict in *Begonia* [J]. Edinburgh Journal of Botany, 79: 1-28.

Clement, W L, Tebbitt, M C, Forrest, L L, et al., 2004. Phylogenetic position and biogeography of *Hillebrandia sandwicensis* (Begoniaceae): a rare Hawaiian relict [J]. American Journal of Botany, 91(6): 905-917.

Ding, Y, Zhang, W, et al., 2017. Cultivation and appreciation of wild begonias [M]. Nanjing: Phoenix Science Press. [丁友芳, 张万旗, 等, 2017. 野生秋海棠的引种栽培与鉴赏 [M]. 南京: 江苏凤凰科学技术出版社.]

Doorenbos, J, Sosef, M S M, De Wilde, J J F E, 1998. The sections of *Begonia* including descriptions, keys and species lists (Studies in Begoniaceae VI) [J]. Wageningen Agricultural University Papers, 98 (2): 1-289.

Dong, W, 2023. Review and prospect of *Begonia*. In: Ma, J, eds. *China - Mother of gardens, in the twenty-first century·Vol. 4. chap. 5* [M]. Beijing: China Forestry Publishing House, 275-389. [董文珂, 2023. 秋海棠属: 回顾与展望. 马金双: 中国——二十一世纪的园林之母·第四卷·第5章 [M]. 北京: 中国林业出版社: 275-389.]

Goodall-Copestake, W P, Harris, D J, Hollingsworth, P M, 2009. The origin of a mega-diverse genus: dating *Begonia* (Begoniaceae) using alternative datasets, calibrations and relaxed clock methods [J]. Botanical Journal of the Linnean Society, 159(3): 363-380.

Goodall-Copestake, W P, Pérez-Espona, S, Harris, D J, Hollingsworth, P M, 2010. The early evolution of the mega-diverse genus *Begonia* (Begoniaceae) inferred from organelle DNA phylogenies [J]. Biological Journal of the Linnean Society, 101(2): 243-250.

Gu, C, Peng, C-I, Turland, N J, 2007. Flora of China (OL). http://www.iplant.cn/foc.

Guan, K, Li, J. et al., 2019. Overview of Begonias in China [M]. Beijing: Beijing Publishing Group. [管开云, 李景秀, 等, 2019. 秋海棠属植物纵览 [M]. 北京: 北京出版集团.]

Li, L, Chen, X, Fang, D, et al., 2022. Genomes shed light on the evolution of *Begonia*, a mega-diverse genus [J]. New Phytologist, 234(1): 295-310.

Mark, H, Peng, C-I, 2018. Asian *Begonia*: 300 species portraits [M]. Taiwan: KBCC Press & RBGE.

Moonlight, P W, Ardi, W H, Padilla L A, et al., 2018. Dividing and conquering the fastest–growing genus: towards a natural sectional classification of the mega–diverse genus *Begonia* (Begoniaceae) [J]. Taxon, 67(2): 267-323.

Peng, J, 2018. For Love, Travel to the Ends of the Earth and Find the Begonias [M]. Taibei: My House Publishing Co., Ltd, Cite Publishing Ltd. [彭镜毅, 2018. 为爱走天涯, 踏觅秋海棠 [M]. 台北: 城邦文化事业股份有限公司, 麦浩斯出版 .]

Rajbhandary, S, Hughes, M, Phutthai, T, et al., 2011. Asian *Begonia*: out of Africa via the Himalayas? [J]. Gardens' Bulletin Singapore, 63(1 & 2): 277-286.

Shui, Y, Chen, W, et al., 2017. *Begonia* of China [M]. Kunming: Yunnan Science & Technology Press, Yunnan Publishing Group Corporation. [税玉民, 陈文红, 等, 2017. 中国秋海棠 [M]. 昆明: 云南科技出版社 .]

Shui, Y, Chen, W, Peng, H, et al., 2019. Taxonomy of Begonias [M]. Kunming: Yunnan Science & Technology Press.

Stults, D Z, Axsmith, B J, 2011. First macrofossil record of *Begonia* (Begoniaceae) [J]. American Journal of Botany, 98(1): 150-153.

Tebbitt, M C, 2005. Begonias: Cultivation, Identification, and Natural History [M]. Portland: Timber Press, Inc.

Thomas, D C, 2010. Phylogenetics and historical biogeography of Southeast Asian *Begonia* (Begoniaceae) [D]. Scotland: University of Glasgow.

Tian, D, Wang, W, Dong, L, et al., 2021. A new species (*Begonia giganticaulis*) of Begoniaceae from southern Xizang (Tibet) of China [J]. PhytoKeys, 187: 189-205.

参考网站
Websites

秋海棠科国际数据库 (The International Database of the BEGONIACEAE): http://ibegonias.filemakerstudio.com.au/index.php?-link=Home

美国秋海棠协会 (The American Begonia Society): https://www.begonias.org/

秋海棠资源中心 (Begonia Resource Centre): https://padme.rbge.org.uk/Begonia/home

世界植物在线 (Plants of the World Online): https://powo.science.kew.org/?page=quickSearch&plantName=Begonia

双花秋海棠
Begonia biflora T. C. Ku

方氏秋海棠
Begonia fangii Y. M. Shui & C. I Peng

秋海棠原生种 Species

无苞组
Begonia Sect. *Augustia* (Klotzsch) A. DC.

开普敦秋海棠
Begonia dregei Otto & A. Dietr.

多年生草本，耐旱；具直立茎，主茎基部膨大。叶茎生，形态似枫叶，深裂，无毛，叶片表面绿色，部分植株叶面有白色斑点。花白色；蒴果具近等长3翅；子房3室，中轴胎座；花柱3。

分布： 原产于南非东部。

生境： 海岸附近的森林。

Perennial herb, drought-tolerant; stems erect, with a swollen caudex. Leaves are cauline-shaped, maple-shaped, divided or parted, glabrous, and the adaxial surface is green or with many white dots. Flowers white; capsule with subequal 3 wings; ovary 3-locular, placentae axile; styles 3.

Distribution: Native to Eastern South Africa.

Habitat: In forest near the coast.

A

10cm

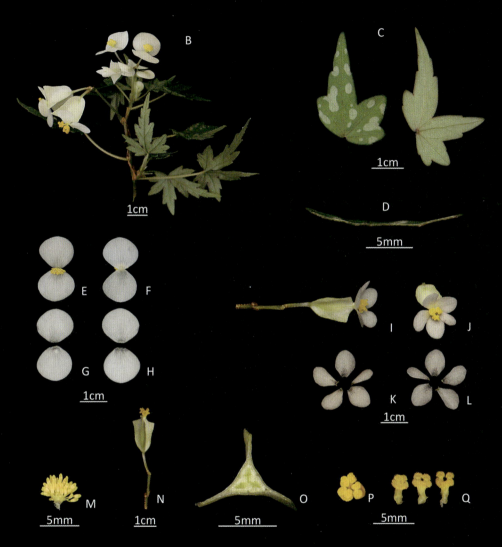

A. 植株　B. 茎　C. 叶片正反面　D. 叶片横切面　E. 雄花正面观　F. 雄花反面观　G. 雄花花被片正面观　H. 雄花花被片反面观　I. 雌花侧面观　J. 雌花正面观　K. 雌花花被片正面观　L. 雌花花被片反面观　M. 雄蕊　N. 蒴果　O. 子房横切面　P. 雌蕊　Q. 花柱和柱头

A. Habit　B. Stem　C. Leaf, adaxial and abaxial view　D. Leaf, transversely sectioned　E. Staminate flower, adaxial view　F. Staminate flower, abaxial view　G. Tepals of staminate flower, adaxial view　H. Tepals of staminate flower, abaxial view　I. Pistillate flower, lateral view　J. Pistillate flower, adaxial view　K. Tepals of pistillate flower, adaxial view　L. Tepals of pistillate flower, abaxial view　M. Androecium, lateral view　N. Capsule　O. Ovary, transversely sectioned　P. Gynoecium, adaxial view　Q. Styles and stigmas

刺莲花秋海棠组 *Begonia* Sect. *Loasibegonia* A. DC.

微籽秋海棠
Begonia microsperma Warb.

多年生草本，高可达20cm；根状茎粗壮。叶片盾状，椭圆形或椭圆卵形至宽卵形；表面密被泡状突起。花淡黄色至橙黄色；蒴果狭椭圆形到狭倒卵形，无翅或具4小翅；子房4室，中轴胎座；花柱4。

分布： 原产于喀麦隆西部。

生境： 潮湿的岩石表面或附生在瀑布旁的榕属植物的树干上。

Perennial herb, up to 20 cm high; rhizomes rather stout. Leaves peltate, elliptic, or elliptic-ovate to broadly; adaxial surface bullate. Flowers yellowish to yellow; capsule narrowly elliptic-oblong to narrowly obovate, with no or very small 4 wings; ovary 4-locular, placentae axile; styles 4.

Distribution: Native to Western Cameroon.

Habitat: Clinging to sheer spray-soaked rock faces or epiphytically on *Ficus* plants near a waterfall.

A

1cm

A. 植株　B. 花序　C. 叶片正反面　D. 叶片横切面　E. 雄花正面观　F. 雄花反面观　G. 雄花花被片正面观　H. 雄花花被片反面观　I. 雌花侧面观　J. 雌花正面观　K. 雌花花被片正面观　L. 雌花花被片反面观　M. 雄蕊　N. 蒴果　O. 子房横切面　P. 雌蕊　Q. 花柱和柱头

A. Habit　B. Inflorescence　C. Leaf, adaxial and abaxial view　D. Leaf, transversely sectioned　E. Staminate flower, adaxial view　F. Staminate flower, abaxial view　G. Tepals of staminate flower, adaxial view　H. Tepals of staminate flower, abaxial view　I. Pistillate flower, lateral view　J. Pistillate flower, adaxial view　K. Tepals of pistillate flower, adaxial view　L. Tepals of pistillate flower, abaxial view　M. Androecium, lateral view　N. Capsule　O. Ovary, transversely sectioned　P. Gynoecium, lateral view　Q. styles and stigmas

棱果秋海棠

Begonia prismatocarpa Hook.

多年生草本，植株矮小，匍匐。叶片卵状长圆形，先端渐尖，浅裂。雌雄花花被片均为2，黄色，有红色条纹；蒴果长卵形，具4小翅；子房4室，中轴胎座；花柱4。

分布： 原产于赤道几内亚（比奥科岛）、喀麦隆、科特迪瓦。

生境： 附生在林下石壁或树上。

Perennial herb, small, prostrate. Leaves ovate-oblong, apex acuminate, lobed. Male and female flowers tepals 2, yellow with red linear dots; capsule elongated-ovoid, with small 4 wings; ovary 4-locular, placentae axile; styles 4.

Distribution: Native to Equatorial Guinea (Bioko Island), Cameroon, and Cote d' Ivoire.

Habitat: Epiphytically on rocks or trees.

A

2cm

A. 植株　B. 花序　C. 叶片正反面　D. 叶片横切面　E. 雄花正面观　F. 雄花反面观　G. 雄花花被片正面观　H. 雄花花被片反面观　I. 雌花正面观　J. 雌花侧面观　K. 雌花花被片正面观　L. 雌花花被片反面观　M. 雄蕊　N. 蒴果　O. 子房横切面　P. 雌蕊　Q. 花柱和柱头

A. Habit　B. Inflorescence　C. Leaf, adaxial and abaxial view　D. Leaf, transversely sectioned　E. Staminate flower, adaxial view　F. Staminate flower, abaxial view　G. Tepals of staminate flower, adaxial view　H. Tepals of staminate flower, abaxial view　I. Pistillate flower, adaxial view　J. Pistillate flower, lateral view　K. Tepals of pistillate flower, adaxial view　L. Tepals of pistillate flower, abaxial view　M. Androecium, lateral view　N. Capsule　O. Ovary, transversely sectioned　P. Gynoecium, adaxial view　Q. Styles and stigmas

宁巴四翅秋海棠

Begonia quadrialata subsp. *nimbaensis* Sosef

多年生草本，根茎类。叶基生，叶片盾形，宽卵形至近圆形，先端渐尖，边缘具疏浅齿，齿尖具长芒。花黄色至橙黄色；蒴果长卵形，被白色长腺毛，具近等长4翅；子房4室，中轴胎座；花柱4。

分布： 原产于几内亚、利比里亚和科特迪瓦边境的宁巴山。

生境： 海拔350～1 600m荫蔽的岩壁上。

Perennial herb, rhizomatous. Leaves basal, peltate, broadly ovate to suborbicular, apex acuminate, margin sparsely toothed, pointed tips with long awns. Flowers yellow to orange-yellow; capsule elongated-ovoid, covered in long white glandular hairs, with subequal 4 wings; ovary 4-locular, placentae axile; styles 4.

Distribution: Native to Mount Nimba, on the border among Guinea, Liberia, and Cote d'Ivoire.

Habitat: Shady rock walls, 350-1,600 m elevation.

A

3cm

A. 植株　B. 花序　C. 叶片正反面和叶柄　D. 叶片横切面　E. 雌花正面观　F. 雌花侧面观　G. 雌花花被片正面观　H. 雌花花被片反面观　I. 雄花侧面观　J. 雄花花被片反面观　K. 雄花花被片正面观　L. 雄蕊　M. 蒴果　N. 子房横切面　O. 雌蕊　P. 花柱和柱头

A. Habit　B. Inflorescence　C. Leaf, adaxial and abaxial view, and petiole　D. Leaf, transversely sectioned　E. Pistillate flower, adaxial view　F. Pistillate flower, lateral view　G. Tepals of pistillate flower, adaxial view　H. Tepals of pistillate flower, abaxial view　I. Staminate flower, lateral view　J. Tepals of staminate flower, abaxial view　K. Tepals of staminate flower, adaxial view　L. Androecium, lateral view　M. Capsule　N. Ovary, transversely sectioned　O. Gynoecium, lateral view　P. Styles and stigmas

鳞叶秋海棠

Begonia scutifolia Hook. f.

多年生草本，具根状茎，匍匐。叶片斜卵形或圆形，边缘具缘毛，不明显锯齿或近全缘，两面无毛。花黄色，雄雌花被片均为2；苞片和小苞片小，卵形，膜质，棕色，具卷曲的长纤毛；蒴果长条形，具4小翅或无翅；子房4室，中轴胎座，花柱4。

分布： 原产于喀麦隆、加蓬、刚果（金）。

生境： 瀑布附近的垂直岩石表面或在长满苔藓的树干上。

Perennial herb, rhizomatous, creeping. Both surfaces are glabrous, leaves obliquely ovate or orbicular, apical acuminate, margin ciliate, inconspicuously serrate or subentire. Flowers yellow, male and female flowers tepals 2, bracts and bracteoles small, ovate, membranous, brown, long cilia with curly hairs; capsule linear, with no or very narrow 4 wings; ovary 4-locular, placentae axile; styles 4.

Distribution: Native to Cameroon, Gabon, and D. R. Congo.

Habitat: On vertical rock faces near waterfalls or on mossy tree trunks.

A
5cm

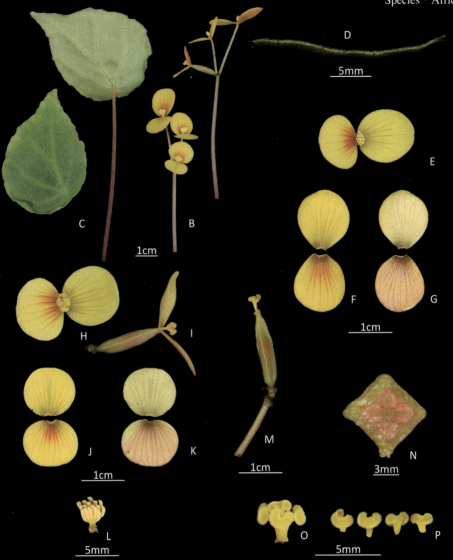

A. 植株　B. 花序　C. 叶片正反面　D. 叶片横切面　E. 雄花正面观　F. 雄花花被片正面观　G. 雄花花被片反面观　H. 雌花正面观　I. 雌花侧面观　J. 雌花花被片正面观　K. 雌花花被片反面观　L. 雄蕊　M. 蒴果　N. 子房横切面　O. 雌蕊　P. 花柱和柱头

A. Habit　B. Inflorescence　C. Leaf, adaxial and abaxial view　D. Leaf, transversely sectioned　E. Staminate flower, adaxial view　F. Tepals of staminate flower, adaxial view　G. Tepals of staminate flower, abaxial view　H. Pistillate flower, adaxial view　I. Pistillate flower, lateral view　J. Tepals of pistillate flower, adaxial view　K. Tepals of pistillate flower, abaxial view　L. Androecium, lateral view　M. Capsule　N. Ovary, transversely sectioned　O. Gynoecium, lateral view　P. Styles and stigmas

非洲四被组 *Begonia* Sect. *Tetraphila* A. DC.

桑寄生状秋海棠
Begonia loranthoides Hook. f.

多年生草本，茎匍匐，稍具木质化。叶片宽卵形，革质，全缘，光滑无毛。花通常呈白色；蒴果长圆筒状，无毛，无翅或近等长4翅；子房4室，中轴胎座；花柱4。

分布：原产于热带西非中西部（圣多美和普林西比、喀麦隆）。

生境：附生在海拔400～1 800m的热带雨林树干上。

Perennial herb, stems prostrate, somewhat woody. Leaves broadly ovate, leathery, margin entire, hairless. Flowers are usually white; capsules long, cylindric, glabrous, wingless, or equally 4-winged; ovary 4-locular, placentae axile; styles 4.

Distribution: Native to West-Central Tropical Africa (Sao Tome & Principe, Cameroon).

Habitat: Epiphytically on trunks in rain forests at *ca*. 400-1,800 m elevation.

A

5cm

3cm

A. 植株　B. 叶片正反面　C. 叶片横切面　D. 雄花侧面观　E. 雄花花被片正面观　F. 雄花花被片反面观　G. 雌花正面观　H. 雌花侧面观　I. 雌花花被片正面观　J. 雌花花被片反面观　K. 雄蕊　L. 蒴果　M. 子房横切面　N. 雌蕊正面观　O. 花柱和柱头

A. Habit　B. Leaf, adaxial and abaxial view　C. Leaf, transversely sectioned
D. Staminate flower, lateral view　E. Tepals of staminate flower, adaxial view
F. Tepals of staminate flower, abaxial view　G. Pistillate flower, abaxial view
H. Pistillate flower, lateral view　I. Tepals of pistillate flower, adaxial view
J. Tepals of pistillate flower, abaxial view　K. Androecium　L. Capsule　M. Ovary,
transversely sectioned　N. Gynoecium, adaxial view　O. Styles and stigmas

匍茎组 *Begonia* Sect. *Kollmannia* Moonlight

红芒秋海棠

Begonia thelmae L. B. Sm. & Wassh.

多年生草本，茎弯曲。叶2列，叶柄纤细，疏生柔毛；叶片椭圆形到长圆形，边缘近规则圆齿，两侧疏生短柔毛。花序腋生，极其纤细；花白色；蒴果具近等长3翅；子房3室，中轴胎座；花柱4。

分布： 原产于巴西（里约热内卢、圣卡塔琳娜）。

生境： 附生在雨林的树干或潮湿的环境。

Perennial herb, stems curved. Leaves 2 rows, petioles slender, sparsely pilose; blades oval to oblong, margin with nearly regular round teeth, two sides sparsely pubescent. Inflorescences axillary, extremely slender; flowers white, capsule with subequal 3 wings; ovary 3-locular, placentae axile; styles 4.

Distribution: Native to Brazil (Rio de Janeiro, Santa Catarina).

Habitat: Epiphytically on tree trunks in the rainforest or wet biome.

A

5cm

　　A. 植株　B. 花序　C. 茎和叶片正反面　D. 叶片横切面　E. 雄花正面观　F. 雄花反面观　G. 雄花花被片正面观　H. 雄花花被片反面观　I. 雌花正面观　J. 雌花侧面观　K. 雌花花被片正面观　L. 雌花花被片反面观　M. 雄蕊　N. 蒴果　O. 子房横切面　P. 雌蕊　Q. 花柱和柱头

　　A. Habit　B. Inflorescence　C. Stem, and leaf, adaxial and abaxial view　D. Leaf, transversely sectioned　E. Staminate flower, adaxial view　F. Staminate flower, abaxial view　G. Tepals of staminate flower, adaxial view　H. Tepals of staminate flower, abaxial view　I. Pistillate flower, adaxial view　J. Pistillate flower, lateral view　K. Tepals of pistillate flower, adaxial view　L. Tepals of pistillate flower, abaxial view　M. Androecium, lateral view　N. Capsule　O. Ovary, transversely sectioned　P. Gynoecium, adaxial view　Q. Styles and stigmas

美洲双瓣组 *Begonia* Sect. *Donaldia* (Klotzsch) A. DC.

榆叶秋海棠

Begonia ulmifolia Willd.

多年生亚灌木，株高可达3 m，茎直立。叶茎生，叶片卵状披针形，被柔毛；表面深绿色或褐绿色，背面褐红色，叶脉凸起。花白色；蒴果具不等长3翅；子房3室，中轴胎座；花柱3。

分布： 原产于巴西、哥伦比亚、圭亚那、秘鲁、特立尼达和多巴哥、委内瑞拉。

生境： 潮湿的林下灌丛中。

Perennial subshrub, up to 3 m tall, stems erect. Leaves cauline, blades ovate-lanceolate, pubescent; adaxial surface deep green or brownish-green; abaxial surface reddish-brown, veins convex. Flowers white; capsule with unequal 3 wings; ovary 3-locular, placentae axile; styles 3.

Distribution: Native to Brazil, Colombia, Guyana, Peru, Trinidad & Tobago, and Venezuela.

Habitat: Moist or wet areas of forests and groves.

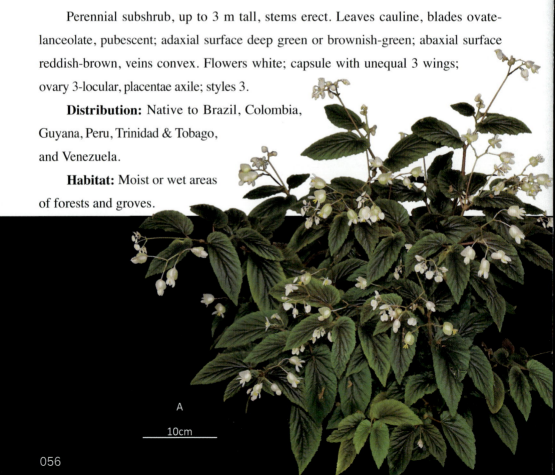

A

10cm

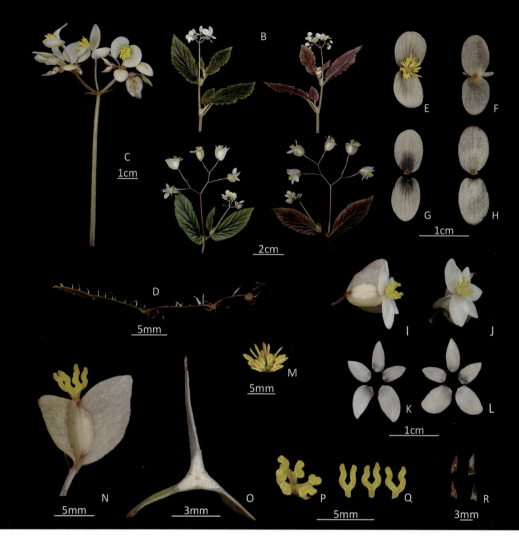

A. 植株　B. 茎，花序腋生　C. 花序　D. 叶片横切面　E. 雄花正面观　F. 雄花反面观　G. 雄花花被片正面观　H. 雄花花被片反面观　I. 雌花侧面观　J. 雌花正面观　K. 雌花花被片正面观　L. 雌花花被片反面观　M. 雄蕊　N. 蒴果　O. 子房横切面　P. 雌蕊　Q. 花柱和柱头　R. 苞片

A. Habit　B. Stem, Inflorescence axillary　C. Inflorescence　D. Leaf, transversely sectioned　E. Staminate flower, adaxial view　F. Staminate flower, abaxial view　G. Tepals of staminate flower, adaxial view　H. Tepals of staminate flower, abaxial view　I. Pistillate flower, lateral view　J. Pistillate flower, adaxial view　K. Tepals of pistillate flower, adaxial view　L. Tepals of pistillate flower, abaxial view　M. Androecium, lateral view　N. Capsule O. Ovary, transversely sectioned　P. Gynoecium, adaxial view　Q. Styles and stigmas R. Bracts, adaxial and abaxial view

巴西秋海棠组 *Begonia* Sect. *Pritzelia* (Klotzsch) A. DC.

柯蒂秋海棠
Begonia curtii L. B. Sm. & B. G. Schub.

多年生草本，高30 ～ 40 cm，茎直立，近肉质，无毛。叶片全缘，椭圆形，不对称；叶柄包被。花白色至粉色；雄花花被片4，雌花花被片5，蒴果近等长3翅，子房3室，中轴胎座；花柱3。

分布：原产于巴西东南部。

生境：不详。

Perennial herb, 30-40 cm tall; stems erect, subfleshy, glabrous. Leaf blades entire margin, elliptic, asymmetrical; petioles coat. Flowers white to pink; male flowers tepals 4, female flowers tepals 5; capsule with subequal 3 wings; ovary 3-locular, placentae axile; styles 3.

Distribution: Native to Southeastern Brazil.

Habitat: Unknown.

A

5cm

A. 植株　B. 花序　C. 叶片正反面　D. 叶片横切面　E. 雄花正面观　F. 雄花反面观　G. 雄花花被片正面观　H. 雄花花被片反面观　I. 雌花正面观　J. 雌花侧面观　K. 雌花花被片正面观　L. 雌花花被片反面观　M. 蒴果　N. 子房横切面　O. 雌蕊　P. 花柱和柱头　Q. 雄蕊

A. Habit　B. Inflorescence　C. Leaf, adaxial and abaxial view　D. Leaf, transversely sectioned　E. Staminate flower, adaxial view　F. Staminate flower, abaxial view　G. Tepals of staminate flower, adaxial view　H. Tepals of staminate flower, abaxial view　I. Pistillate flower, adaxial view　J. Pistillate flower, lateral view　K. Tepals of pistillate flower, adaxial view　L. Tepals of pistillate flower, abaxial view　M. Capsule　N. Ovary, transversely sectioned　O. Gynoecium, adaxial view　P. Styles and stigmas　Q. Androecium, lateral view

迪特里希秋海棠

Begonia dietrichiana Irmsch.

多年生草本，茎直立，红棕色。叶片卵状披针形，无毛；表面深绿色或褐绿色，叶脉周围呈浅绿色，背面红褐色，靠近叶脉处渐渐变成黄绿色。花白色，较小；蒴果具不等长3翅；子房3室，中轴胎座；花柱3。

分布： 原产于巴西东南部。

生境： 潮湿的林下环境。

Perennial herb, stems erect, reddish brown. Leaf blades ovate-lanceolate, glabrous; abaxial surface dark green or brownish green, becoming pale green near venation, abaxial surface reddish brown, becoming greenish yellow near venation. Flowers white, small; capsule with unequal 3 wings; ovary 3-locular, placentae axile; styles 3.

Distribution: Native to Southeastern Brazil.

Habitat: In moist or wet areas of forests.

A
10cm

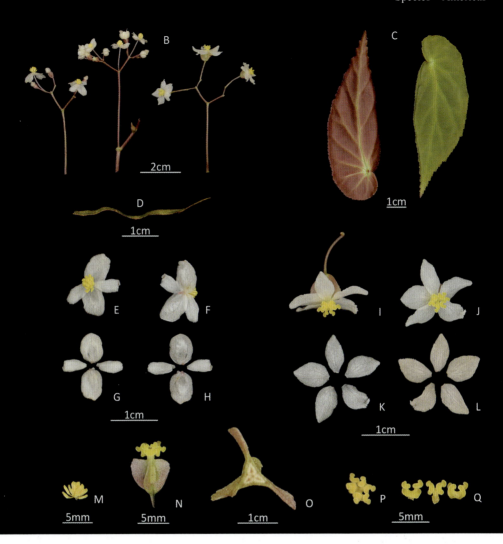

　　A. 植株　B. 花序　C. 叶片正反面　D. 叶片横切面　E. 雄花正面观　F. 雄花反面观　G. 雄花花被片正面观　H. 雄花花被片反面观　I. 雌花侧面观　J. 雌花正面观　K. 雌花花被片正面观　L. 雌花花被片反面观　M. 雄蕊　N. 蒴果　O. 子房横切面　P. 雌蕊　Q. 花柱和柱头

　　A. Habit　B. Inflorescence　C. Leaf, adaxial and abaxial view　D. Leaf, transversely sectioned　E. Staminate flower, adaxial view　F. Staminate flower, abaxial view　G. Tepals of staminate flower, adaxial view　H. Tepals of staminate flower, abaxial view　I. Pistillate flower, lateral view　J. Pistillate flower, adaxial view　K. Tepals of pistillate flower, adaxial view　L. Tepals of pistillate flower, abaxial view　M. Androecium, lateral view　N. Capsule O. Ovary, transversely sectioned　P. Gynoecium, adaxial view　Q. Styles and stigmas

刺萼秋海棠

Begonia echinosepala Regel

亚灌木状多年生草本，茎直立，稍纤细。叶片披针形，具不规则细锯齿，深绿色。花白色，疏或中度被柔毛；蒴果具不等长3翅；子房3室，中轴胎座；花柱3。

分布： 原产于巴西东南部。

生境： 山谷底部、斜坡以及雨水冲刷较少的岩石上、靠近小溪的岩石区域。

Perennial subshrub herb, stems erect, slightly slender. Leaves lanceolate, margin irregularly serrulate, dark green. Flowers white, sparsely to moderately pubescent; capsule with unequal 3 wings; ovary 3-locular, placentae axile; styles 3.

Distribution: Native to Southeastern Brazil.

Habitat: In the bottom of valleys, on slopes and rocks, with little exposure to rainwater runoff, and in the rocky zones near small streams.

A

10cm

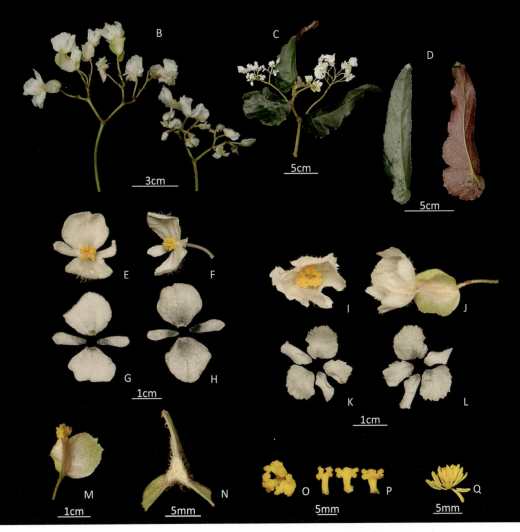

　　A. 植株　B. 花序　C. 茎，花序腋生　D. 叶片正反面　E. 雄花正面观　F. 雄花侧面观　G. 雄花花被片正面观　H. 雄花花被片反面观　I. 雌花正面观　J. 雌花侧面观　K. 雌花花被片正面观　L. 雌花花被片反面观　M. 蒴果　N. 子房横切面　O. 雌蕊　P. 花柱和柱头　Q. 雄蕊

　　A. Habit　B. Inflorescence　C. Stem, inflorescence axillary　D. Leaf, adaxial and abaxial view　E. Staminate flower, adaxial view　F. Staminate flower, lateral view　G. Tepals of staminate flower, adaxial view　H. Tepals of staminate flower, abaxial view　I. Pistillate flower, adaxial view　J. Pistillate flower, lateral view　K. Tepals of pistillate flower, adaxial view　L. Tepals of pistillate flower, abaxial view　M. Capsule　N. Ovary, transversely sectioned　O. Gynoecium, adaxial view　P. Styles and stigmas　Q. Androecium, lateral view

金线秋海棠

Begonia listada L. B. Sm. & Wassh.

多年生草本，茎直立或近直立，密被缘毛。叶片近全缘，疏生柔毛；表面深绿色，沿主轴有一突出的白色条纹，形似猫眼，背面红色。花序腋生，被柔毛；花白色或浅红色；蒴果椭圆形，具略不等长3翅；子房3室，中轴胎座；花柱3。

分布：原产于巴拉圭。

生境：热带雨林潮湿荫蔽处。

Perennial herb, stems erect or suberect, densely ciliate. Leaf blades margin subentire, pubescent sparsely throughout margin; adaxial surface dark green, with a prominent white stripe along main axis, like cat's eyes; abaxial surface red. Inflorescences peduncle axillary, pubescent; flowers white or white reddish; capsule elliptic, with slightly unequal 3 wings; ovary 3-locular, placentae axile; styles 3.

Distribution: Native to Paraguay.

Habitat: Moist, shady spots in rainforests.

A

10cm

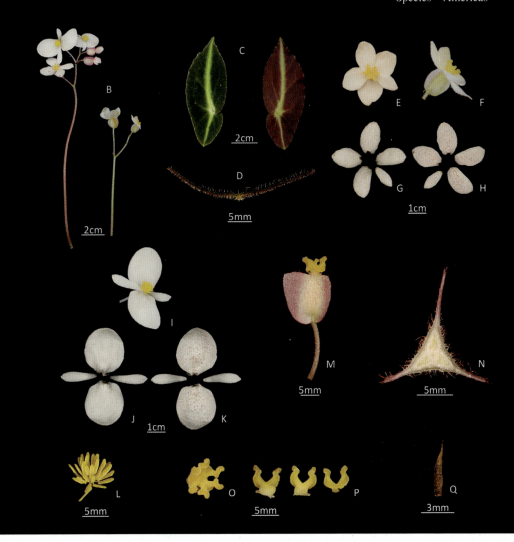

A. 植株　B. 花序　C. 叶片正反面　D. 叶片横切面　E. 雌花正面观　F. 雌花侧面观
G. 雌花花被片正面观　H. 雌花花被片反面观　I. 雄花正面观　J. 雄花花被片正面观
K. 雄花花被片反面观　L. 雄蕊　M. 蒴果　N. 子房横切面　O. 雌蕊　P. 花柱和柱头
Q. 苞片

A. Habit　B. Inflorescence　C. Leaf, adaxial and abaxial view　D. Leaf, transversely
sectioned　E. Pistillate flower, adaxial view　F. Pistillate flower, lateral view　G. Tepals
of pistillate flower, adaxial view　H. Tepals of pistillate flower, abaxial view　I. Staminate
flower, adaxial view　J. Tepals of staminate flower, adaxial view　K. Tepals of staminate
flower, abaxial view　L. Androecium, lateral view　M. Capsule　N. Ovary, transversely
sectioned　O. Gynoecium, adaxial view　P. Styles and stigmas　Q. Bract, adaxial view

奥尔森秋海棠

Begonia olsoniae L. B. Sm. & B. G. Schub.

多年生草本，具匍匐茎。叶片扁椭圆形或心形，全缘，具缘毛；表面铜绿色，背面红色，两面均被毛，叶脉淡绿色。花白色；雌花具小苞片；蒴果卵形，具极不等长3翅；子房3室，中轴胎座；花柱3。

分布：原产于巴西（里约热内卢、圣卡塔琳娜）。

生境：湿热的环境。

Perennial herb, stoloniferous. Leaf blades oblong or cordate, margin whole, ciliated; adaxial surface copper-green, abaxial surface red, both sides covered with hair, veins light green. Flowers white; female flowers bracteolate; capsule ovate, with highly unequal 3 wings; ovary 3-locular, placentae axile; styles 3.

Distribution: Native to Brazil (Rio de Janeiro, Santa Catarina).

Habitat: In wet tropical biome.

A

5cm

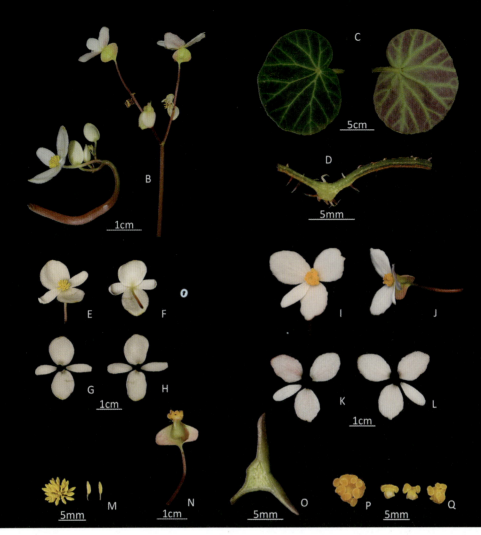

A. 植株　B. 花序　C. 叶片正反面　D. 叶片横切面　E. 雄花正面观　F. 雄花反面观 G. 雄花花被片正面观　H. 雄花花被片反面观　I. 雌花正面观　J. 雌花侧面观　K. 雌花花被片反面观　L. 雌花花被片正面观　M. 雄蕊　N. 蒴果　O. 子房横切面　P. 雌蕊　Q. 花柱和柱头

A. Habit　B. Inflorescence　C. Leaf, adaxial and abaxial view　D. Leaf, transversely sectioned　E. Staminate flower, adaxial view　F. Staminate flower, abaxial view　G. Tepals of staminate flower, adaxial view　H. Tepals of staminate flower, abaxial view　I. Pistillate flower, adaxial view　J. Pistillate flower, lateral view　K. Tepals of pistillate flower, abaxial view　L. Tepals of pistillate flower, adaxial view　M. Androecium, adaxial view　N. Capsule O. Ovary, transversely sectioned　P. Gynoecium, adaxial view　Q. Styles and stigmas

巴西变色秋海棠
Begonia solimutata L. B. Sm. & Wassh.

多年生草本，匍匐，株高可达30 cm。叶互生，叶片近圆形，边缘浅裂；表面深绿色具泡状突起，沿叶脉带黄绿色条纹，边缘疏被红色柔毛，背面红色，沿脉浅绿色。花白色；蒴果亚卵形，具不等长3翅；子房3室，中轴胎座；花柱3。

分布：原产于巴西北部。

生境：潮湿的林下环境。

Perennial herb, prostrate, up to *ca.* 30 cm tall. Leaves alternate, blades suborbicular, margin shallowly lobed; adaxial surface dark green with pustules, stripes along leaf veins yellow-green, margin sparsely red villous, abaxial surface red, light green along veins. Flowers white; capsule subovate, with unequal 3 wings; ovary 3-locular, placentae axile; styles 3.

Distribution: Native to Northern Brazil.

Habitat: On moist environment in forests.

A

5cm

A. 植株　B. 花序　C. 叶片正反面　D. 叶片横切面　E. 雄花正面观　F. 雄花侧面观
G. 雄花花被片正面观　H. 雄花花被片反面观　I. 雌花正面观　J. 雌花侧面观　K. 雌
花花被片正面观　L. 雌花花被片反面观　M. 雄蕊　N. 蒴果　O. 子房横切面　P. 雌
蕊　Q. 花柱和柱头

A. Habit　B. Inflorescence　C. Leaf, adaxial and abaxial view　D. Leaf, transversely
sectioned　E. Staminate flower, adaxial view　F. Staminate flower, lateral view　G. Tepals
of staminate flower, adaxial view　H. Tepals of staminate flower, abaxial view　I. Pistillate
flower, adaxial view　J. Pistillate flower, lateral view　K. Tepals of pistillate flower, adaxial
view　L. Tepals of pistillate flower, abaxial view　M. Androecium, lateral view　N. Capsule
O. Ovary, transversely sectioned　P. Gynoecium, adaxial view　Q. Styles and stigmas

四季秋海棠组 *Begonia* Sect. *Ephemera* Moonlight

软茎秋海棠

Begonia mollicaulis Irmsch.

多年生大草本，茎直立，稍弯曲，具毛。叶纸质，斜卵形，边缘具齿；表面淡绿色，被毛，背面叶脉被毛。聚伞花序，多花，花梗被毛；苞片宿存；雄花花被片4，雌花花被片5；蒴果三角椭圆形，具不等长3翅；子房3室，中轴胎座；花柱3。

分布：原产于巴西南部。

生境：湿热的环境。

Perennial large herb, stems erect, slightly curved, hairy. Leaf blades papery, obliquely ovate, margin with tooth; adaxial surface light green, hairy, abaxial surface hairy on veins. Inflorescences cymiferous, floriferous, peduncle hairy; bracts persistent; male flowers tepals 4, female flowers tepals 5; capsule oblong trigonal, with unequal 3 wings; ovary 3-locular, placentae axile; styles 3.

Distribution: Native to Southern Brazil.

Habitat: In wet tropical biome.

A

10cm

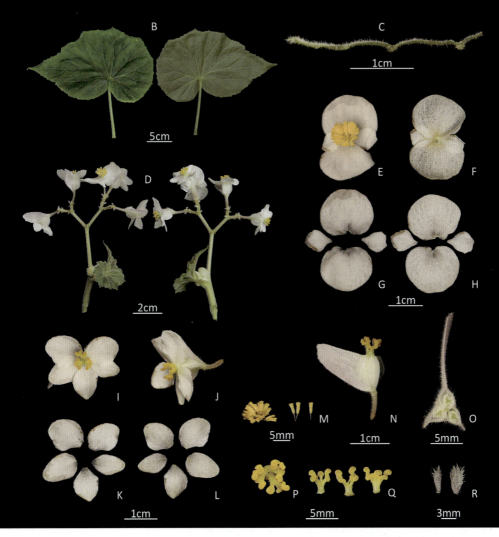

A. 植株　B. 叶片正反面　C. 叶片横切面　D. 花序　E. 雄花正面观　F. 雄花反面观 G. 雄花花被片正面观　H. 雄花花被片反面观　I. 雌花正面观　J. 雌花侧面观　K. 雌花花被片正面观　L. 雌花花被片反面观　M. 雄蕊　N. 蒴果　O. 子房横切面　P. 雌蕊　Q. 花柱和柱头　R. 苞片

A. Habit　B. Leaf, adaxial and abaxial view　C. Leaf, transversely sectioned D. Inflorescence　E. Staminate flower, adaxial view　F. Staminate flower, abaxial view G. Tepals of staminate flower, adaxial view　H. Tepals of staminate flower, abaxial view I. Pistillate flower, adaxial view　J. Pistillate flower, lateral view　K. Tepals of pistillate flower, adaxial view　L. Tepals of pistillate flower, abaxial view　M. Androecium, lateral view　N. Capsule　O. Ovary, transversely sectioned　P. Gynoecium, adaxial view Q. Styles and stigmas　R. Bracts, adaxial view

竹节秋海棠组 *Begonia* Sect. *Gaerdtia* (Klotzsch) A. DC.

斑叶竹节秋海棠
Begonia maculata Raddi

亚灌木类，呈竹节状，高约1.5 m。叶茎生，叶片长椭圆形至卵形，边缘呈波状，表面呈亮绿色，具有圆形银白色斑点，背面呈红色。总状花序，有4～5个分枝，花量大；花白色至粉红色；蒴果白色，无毛，具等长或近等长3翅；子房3室，中轴胎座；花柱3。

分布：原产于巴西东南部。

生境：热带雨林中。

Subshrub, bamboo-like, up to *ca.* 1.5 m tall. Leaves cauline, blades oblong to ovate, margin wavy, adaxial surface shiny green, with rounded silver spots, abaxial surface red. Inflorescences 4-5-branched cymes, floriferous; flowers white to pink; capsule white, glabrous, with equal or subequal 3 wings; ovary 3-locular, placentae axile; styles 3.

Distribution: Native to Southeastern Brazil.

Habitat: In tropical rainforests.

A

20cm

A. 植株　B. 花序　C.叶片正反面　D. 雄花正面观　E. 雄花反面观　F. 雄花花被片正面观　G. 雄花花被片反面观　H. 雌花正面观　I. 雌花侧面观　J. 雌花花被片正面观　K. 雌花花被片反面观　L. 雄蕊　M. 蒴果　N. 子房横切面　O. 雌蕊　P. 花柱和柱头

A. Habit　B. Inflorescence　C. Leaf, adaxial and abaxial view　D. Staminate flower, adaxial view　E. Staminate flower, abaxial view　F. Tepals of staminate flower, adaxial view G. Tepals of staminate flower, abaxial view　H. Pistillate flower, adaxial view　I. Pistillate flower, lateral view　J. Tepals of pistillate flower, adaxial view　K. Tepals of pistillate flower, abaxial view　L. Androecium, lateral view　M. Capsule　N. Ovary, transversely sectioned O. Gynoecium, adaxial view　P. Styles and stigmas

根茎秋海棠组 *Begonia* Sect. *Gireoudia* (Klotzsch) A. DC.

白芷叶秋海棠

Begonia heracleifolia Schltdl. & Cham.

多年生草本，根状茎伸长。叶片近对称，圆形或心形，深裂；边缘波状，具小齿和缘毛；表面墨绿色，沿脉具白色条纹，背面叶脉突出。聚伞状花序，不对称着生于花梗；花白色或粉色；蒴果亚卵形，具不等长3翅；子房3室，中轴胎座；花柱3。

分布：原产于墨西哥至洪都拉斯。

生境：石灰岩山坡上或分布在多种类型的森林，包括高山中等湿度森林、热带半落叶林和橡树林中。

Perennial herb, rhizomes elongate. Leaf blades sub-symmetrical, rounded or cordate, deeply lobed; margin undulate, denticulate and ciliated; abaxial surface dark green with white stripes along the veins, abaxial surface veins protruding. Inflorescences cymiferous, unevenly developed; flowers white or pinkish; capsule subovate, with unequal 3 wings; ovary 3-locular, placentae axile; styles 3.

Distribution: Native from Mexico to Honduras.

Habitat: On limestone slopes or distributed in various types of forests, including montane mesophytic, tropical subdeciduous, and oak forests.

A

10cm

A. 植株　B. 花序　C. 叶片正反面　D. 叶片横切面　E, F. 雄花侧面观　G. 雄花花被片正面观　H. 雄花花被片反面观　I, J. 雌花侧面观　K. 雌花花被片正面观 L. 雌花花被片反面观　M. 雄蕊　N. 蒴果　O. 子房横切面　P. 雌蕊　Q. 花柱和柱头　R. 苞片

A. Habit　B. Inflorescence　C. Leaf, adaxial and abaxial view　D. Leaf, transversely sectioned　E, F. Staminate flower, lateral view　G. Tepals of staminate flower, adaxial view H. Tepals of staminate flower, abaxial view　I, J. Pistillate flower, lateral view　K. Tepals of pistillate flower, adaxial view　L. Tepals of pistillate flower, abaxial view　M. Androecium, lateral view　N. Capsule　O. Ovary, transversely sectioned　P. Gynoecium, adaxial view Q. Styles and stigmas　R. Bracts, adaxial and abaxial view

褐脉秋海棠

Begonia glandulosa A. DC. ex Hook.

多年生草本，根状茎粗壮，向上弯曲，鳞片覆盖。叶柄亮红色，具毛；叶片卵形或心形，或几乎圆形，边缘波状具齿；表面淡绿色，叶脉褐色。花梗纤细，深红色分枝；花多而小，淡绿色或白色；蒴果三角形，具不等长3翅；子房3室，中轴胎座；花柱3。

分布：原产于墨西哥。

生境：云雾缭绕的森林、热带半落叶林或橡树林中，通常出现在岩石坡上，海拔高度450～1 600 m。

Perennial herb, rhizomes stout, curved upward, covered by scales. Petioles bright red and hairy; leaf blades ovate or cordate, or almost rounded, margin undulate teeth; adaxial surface pale green, veins brownish. Peduncle slender, deep red branches and pedicels; flowers numerous and small, light green or white; capsule triangular, with unequal 3 wings; ovary 3-locular, placentae axile; styles 3.

Distribution: Native to Mexico.

Habitat: Cloud forests, tropical sub-deciduous forests, and oak forests, often found on rocky slopes, 450-1,600 m elevation.

A

5cm

A. 植株　B. 花序　C. 叶片正反面　D. 雄花正面观　E. 雄花反面观　F. 雄花花被片正面观　G. 雄花花被片反面观　H. 雌花侧面观　I. 雌花花被片正面观　J. 雌花花被片反面观　K. 雄蕊　L. 蒴果　M. 子房横切面　N. 雌蕊　O. 花柱和柱头　P. 苞片

A. Habit　B. Inflorescence　C. Leaf, adaxial and abaxial view　D. Staminate flower, adaxial view　E. Staminate flower, abaxial view　F. Tepals of staminate flower, adaxial view　G. Tepals of staminate flower, abaxial view　H. Pistillate flower, lateral view　I. Tepals of pistillate flower, adaxial view　J. Tepals of pistillate flower, abaxial view　K. Androecium, adaxial view　L. Capsule　M. Ovary, transversely sectioned　N. Gynoecium, adaxial view　O. Styles and stigmas　P. Bracts, abaxial view

荷叶秋海棠

Begonia nelumbiifolia Schltdl. & Cham.

多年生草本，具根状茎。叶基生，叶片盾状，卵形，似荷叶，先端渐尖；表面绿色，光滑无毛；部分幼叶叶脉呈红色。花白色；蒴果具不等长3翅；子房3室，中轴胎座；花柱3。

分布：原产于墨西哥中部至哥伦比亚。

生境：林下阴湿的沟谷。

Perennial herb, rhizomatous. Leaves basal, blades peltate, ovate, resembling lotus leaves, apex acuminate; adaxial surface green, glabrous; partial young leaves produce red veins. Flowers white; capsule with unequal 3 wings; ovary 3-locular, placentae axile; styles 3.

Distribution: Native from Central Mexico to Colombia.

Habitat: In shady and moist valleys under the forest canopy.

A

10cm

  A. 植株 B. 花序 C. 叶片正反面 D. 叶片横切面 E. 雄花侧面观 F. 雄花反面观 G. 雄花花被片正面观 H. 雄花花被片反面观 I. 雌花正面观 J. 雌花侧面观 K. 雌花花被片正面观 L. 雌花花被片反面观 M. 雄蕊 N. 蒴果 O. 子房横切面 P. 雌蕊 Q. 花柱和柱头 R. 苞片

  A. Habit B. Inflorescence C. Leaf, adaxial and abaxial view D. Leaf, transversely sectioned E. Staminate flower, lateral view F. Staminate flower, abaxial view G. Tepals of staminate flower, adaxial view H. Tepals of staminate flower, abaxial view I. Pistillate flower, adaxial view J. Pistillate flower, lateral view K. Tepals of pistillate flower, adaxial view L. Tepals of pistillate flower, abaxial view M. Androecium, lateral view N. Capsule O. Ovary, transversely sectioned P. Gynoecium, adaxial view Q. Styles and stigmas R. Bracts, adaxial view

列梅秋海棠

Begonia thiemei A. DC.

多年生草本，具根状茎。叶基生，掌状复叶，小叶披针形，边缘具不规则锯齿；表面绿色，光滑无毛，背面棕色。花白色至绿色；蒴果卵球形，具不等长3翅；子房3室，中轴胎座；花柱3。

分布： 原产于墨西哥南部、危地马拉、洪都拉斯。

生境： 林下岩石上。

Perennial herb, rhizomatous. Leaves basal, palmately compound, blades lanceolate, margin irregularly serrate; adaxial surface green, glabrous, abaxial surface brownish. Flowers white to green; capsule oval, with unequal 3 wings; ovary 3-locular, placentae axile; styles 3.

Distribution: Native to Southern Mexico, Guatemala, and Honduras.

Habitat: On rocks in forests.

A

10cm

A. 植株　B. 花序　C. 叶片正反面　D. 雄花正面观　E. 雄花反面观　F. 雄花花被片正面观　G. 雄花花被片反面观　H. 雌花侧面观　I. 雌花正面观　J. 雌花花被片正面观　K. 雌花花被片反面观　L. 雄蕊　M. 蒴果　N. 子房横切面　O. 雌蕊　P. 花柱和柱头　Q. 苞片

A. Habit　B. Inflorescence　C. Leaf, adaxial and abaxial view　D. Staminate flower, adaxial view　E. Staminate flower, abaxial view　F. Tepals of staminate flower, adaxial view G. Tepals of staminate flower, abaxial view　H. Pistillate flower, lateral view　I. Pistillate flower, adaxial view　J. Tepals of pistillate flower, adaxial view　K. Tepals of pistillate flower, abaxial view　L. Androecium, lateral view　M. Capsule　N. Ovary, transversely sectioned　O. Gynoecium, adaxial view　P. Styles and stigmas　Q. Bract, adaxial view

列梅秋海棠

Begonia thiemei A. DC.

A

10cm

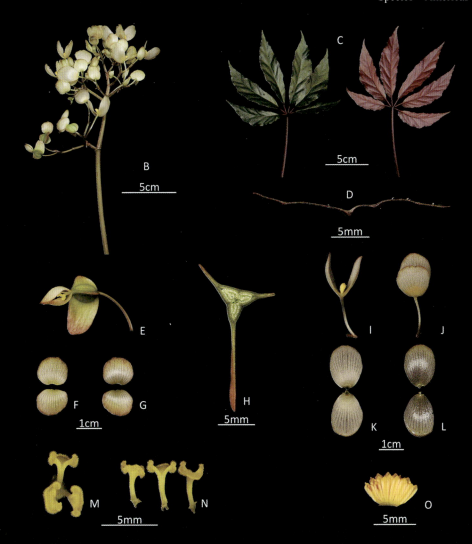

A. 植株　B. 花序　C. 叶片正反面　D. 叶片横切面　E. 雌花侧面观　F. 雌花花被片正面观　G. 雌花花被片反面观　H. 子房横切面　I，J. 雄花侧面观　K. 雄花花被片正面观　L. 雄花花被片反面观　M. 雌蕊　N. 花柱和柱头　O. 雄蕊

A. Habit　B. Inflorescence　C. Leaf, adaxial and abaxial view　D. Leaf, transversely sectioned　E. Pistillate flower, lateral view　F. Tepals of pistillate flower, adaxial view G. Tepals of pistillate flower, abaxial view　H. Ovary, transversely sectioned　I, J. Staminate flower, lateral view　K. Tepals of staminate flower, adaxial view　L. Tepals of staminate flower, abaxial view　M. Gynoecium, adaxial view　N. Styles and stigmas　O. Androecium, lateral view

茄蕊组 *Begonia* Sect. *Solananthera* A. DC.

气根秋海棠

Begonia radicans Vell.

多年生草本，攀爬植物，长度可达3～5 m，茎无毛。叶片稍不对称，卵形或椭圆形，先端渐尖，基部稍偏斜，边缘平至微波状，两面无毛；表面绿色，背面浅绿色。雄花花被片4，雌花5；花被片颜色从基部向外缘逐渐由红色渐变为白色；蒴果具有小腺毛；蒴果具微小腺毛，具不等3翅；子房3室，中轴胎座；花柱3。

分布： 原产于巴西南部及东南部。

生境： 热带雨林树上。

Perennial herb, climbing plant, up to 3-5 m long, stems glabrous. Leaf blades slightly asymmetrical, ovate or elliptic, apex acuminate, base slightly oblique, margin flat to slightly undulate, both sides glabrous; adaxial surface green, abaxial surface light green. Male flowers tepals 4, female flowers tepals 5; color of tepals transitions gradually from red at the base to white towards the outer edges; capsule, tiny glandular hairs, with unequal 3 wings; ovary 3-locular, placentae axile; styles 3.

Distribution: Native to Southern and Southeastern Brazil.

Habitat: On trees in the Selva tropical forest.

10cm

A

5cm

　　A. 植株　B. 花序　C. 叶片正反面　D. 雄花正面观　E. 雄花侧面观　F. 雄花花被片正面观　G. 雄花花被片反面观　H. 雌花正面观　I. 雌花侧面观　J. 雌花花被片正面观　K. 雌花花被片反面观　L. 雄蕊　M. 蒴果　N. 子房横切面　O. 雌蕊　P. 花柱和柱头

　　A. Habit　B. Inflorescence　C. Leaf, adaxial and abaxial view　D. Staminate flower, adaxial view　E. Staminate flower, lateral view　F. Tepals of staminate flower, adaxial view　G. Tepals of staminate flower, abaxial view　H. Pistillate flower, adaxial view　I. Pistillate flower, lateral view　J. Tepals of pistillate flower, adaxial view　K. Tepals of pistillate flower, abaxial view　L. Androecium, adaxial and lateral view　M. Capsule　N. Ovary, transversely sectioned　O. Gynoecium, adaxial view　P. Styles and stigmas

美洲盾叶组 *Begonia* Sect. *Pereira* Brade

柳伯斯秋海棠

Begonia lubbersii É. Morren

半灌木，具分枝，茎顶端下垂。叶片呈长方菱形，狭窄，边缘近波状，无毛；叶片表面深绿色，发亮，有零星或小片的银色斑点，背面呈紫色。聚伞花序腋生，下垂；花大，花被片绿色、粉红色或白色，具有持久的花香；蒴果具不等长3翅；子房3室，中轴胎座；花柱3。

分布： 原产于巴西东南部。

生境： 湿热环境，最早发现生长于树蕨植物树干上，耐热耐湿、耐晒也耐阴。

Subshrub, branched, stems apex sag. Leaf blades oblong rhomboid, narrow, margin sub-undulate, glabrous; adaxial surface dark green, shiny, with infrequent or confluent silver spots; abaxial surface rather purple. Cyme axillary, nodding; flowers big, green, pink or white, lasting fragrance; capsule with unequal 3 wings; ovary 3-locular, placentae axile; styles 3.

Distribution: Native to Southeast Brazil.

Habitat: In wet tropical biome, firstly discovered growing from the stem of a tree fern, resistant to humidity and heat, sun-proof or shade-tolerant.

A

5cm

A. 植株　B. 花序　C. 叶片正反面　D. 叶片横切面　E. 雄花正面观　F. 雄花侧面观　G. 雄花花被片正面观　H. 雄花花被片反面观　I. 雌花正面观　J. 雌花侧面观　K. 雌花花被片正面观　L. 雌花花被片反面观　M. 雄蕊　N. 蒴果　O. 子房横切面　P. 雌蕊　Q. 花柱和柱头

A. Habit　B. Inflorescence　C. Leaf, adaxial and abaxial view　D. Leaf, transversely sectioned　E. Staminate flower, adaxial view　F. Staminate flower, lateral view　G. Tepals of staminate flower, adaxial view　H. Tepals of staminate flower, abaxial view　I. Pistillate flower, adaxial view　J. Pistillate flower, lateral view　K. Tepals of pistillate flower, adaxial view　L. Tepals of pistillate flower, abaxial view　M. Androecium, lateral view　N. Capsule O. Ovary, transversely sectioned　P. Gynoecium, adaxial view　Q. Styles and stigmas

丛茎组 *Begonia* Sect. *Lepsia* (Klotzsch) A. DC.

多叶秋海棠
Begonia foliosa Kunth

多年生草本，茎直立。叶片茎生，细小；叶片卵形，边缘具不规则松散锯齿；表面深绿色，背面浅绿色。花白色；蒴果具不等长3翅；子房3室，中轴胎座；花柱3。

分布： 原产于哥伦比亚、厄瓜多尔、秘鲁、委内瑞拉。

生境： 潮湿的山地森林中。

Perennial herb, stems erect. Leaves cauline, small; blades ovate, margin irregularly loosely serrate; adaxial surface dark green, abaxial surface pale green. Flowers white; capsule with unequal 3 wings; ovary 3-locular, placentae axile; styles 3.

Distribution: Native to Colombia, Ecuador, Peru, and Venezuela.

Habitat: In moist montane forests.

A

5cm

A. 植株　B. 茎，花序腋生　C. 花序　D. 叶片正反面　E. 叶片横切面　F. 雄花正面观 G. 雄花反面观　H. 雄花花被片正面观　I. 雄花花被片反面观　J. 雌花侧面观　K. 雌花正面观　L. 雌花花被片正面观　M. 雌花花被片反面观　N. 雄蕊　O. 子房横切面 P. 雌蕊　Q. 花柱和柱头　R. 蒴果　S. 苞片

A. Habit　B. Stem, inflorescence axillary　C. Inflorescence　D. Leaf, adaxial and abaxial view　E. Leaf, transversely sectioned　F. Staminate flower, adaxial view G. Staminate flower, abaxial view　H. Tepals of staminate flower, adaxial view　I. Tepals of staminate flower, abaxial view　J. Pistillate flower, lateral view　K. Pistillate flower, adaxial view　L. Tepals of pistillate flower, adaxial view　M. Tepals of pistillate flower, abaxial view N. Androecium, lateral view　O. Ovary, transversely sectioned　P. Gynoecium, adaxial view Q. Styles and stigmas　R. Capsule　S. Bracts, adaxial and abaxial view

矛果组 *Begonia* Sect. *Doratometra* (Klotzsch) A. DC.

矮生秋海棠
Begonia humilis Aiton

多年生草本，茎直立，植株矮小。叶互生，叶片不对称，披针形，边缘重锯齿；表面绿色，无毛至疏被柔毛，背面淡绿色至红色，无毛。花白色；蒴果卵球形，无毛，具不等长3翅；子房3室，中轴胎座；花柱3。

分布：原产于玻利维亚、巴西、哥伦比亚、厄瓜多尔、圭亚那、牙买加、秘鲁、苏里南、特立尼达和多巴哥、委内瑞拉。

生境：潮湿阴凉的河岸。

Perennial herb, stems erect, dwarf. Leaves alternate, blades asymmetric, lanceolate, margin double serrate; adaxial surface green, glabrous to sparsely pilose, abaxial surface pale green to red, glabrous. Flowers white; capsule ovoid, glabrous, with unequal 3 wings; ovary 3-locular, placentae axile; styles 3.

Distribution: Native to Bolivia, Brazil, Colombia, Ecuador, Guyana, Jamaica, Peru, Suriname, Trinidad & Tobago, and Venezuela.

Habitat: In moist, shaded banks.

A

5cm

　　A. 植株　B. 茎　C. 花序　D. 叶片正反面　E. 叶片横切面　F. 雄花正面观　G. 雄花花被片正面观　H. 雄花花被片反面观　I. 雌花正面观　J. 雌花侧面观　K. 雌花花被片正面观　L. 雌花花被片反面观　M. 雄蕊　N. 蒴果　O. 子房横切面　P. 雌蕊　Q. 花柱和柱头

　　A. Habit　B. Stem　C. Inflorescence　D. Leaf, adaxial and abaxial view　E. Leaf, transversely sectioned　F. Staminate flower, adaxial view　G. Tepals of staminate flower, adaxial view　H. Tepals of staminate flower, abaxial view　I. Pistillate flower, adaxial view　J. Pistillate flower, lateral view　K. Tepals of pistillate flower, adaxial view　L. Tepals of pistillate flower, abaxial view　M. Androecium, lateral view　N. Capsule　O. Ovary, transversely sectioned　P. Gynoecium, adaxial view　Q. Styles and stigmas

瓦氏秋海棠

Begonia wallichiana Lehm.

一年生具茎植物，通常密被柔毛。叶片斜椭圆形，淡绿色，密被细长的腺毛。花序腋生；花白色；蒴果长圆形、广椭圆形或圆形，具不等长3翅；子房3室，中轴胎座；花柱3。

分布：原产于墨西哥至危地马拉。

生境：季节性干燥的热带环境。

Annual stemmed plants, usually with dense hairs. Leaf blades oblique, oval, and light green, with thick, slender glandular hairs. Inflorescences axillary; flowers white; capsule oblong, oval or round, with unequal 3 wings; ovary 3-locular, placentae axile; styles 3.

Distribution: Native from Mexico to Guatemala.

Habitat: In the seasonally dry tropical biome.

A. 植株　B. 花序　C. 叶片正反面　D. 叶片横切面　E. 雄花侧面观　F. 雄花反面观　G. 雄花花被片正面观　H. 雄花花被片反面观　I. 雌花侧面观　J. 雄花正面观　K. 雌花反面观　L. 雌花花被片正面观　M. 雌花花被片反面观　N. 蒴果　O. 子房横切面　P. 雌蕊　Q. 花柱和柱头　R. 雄蕊

A. Habit　B. Inflorescence　C. Leaf, adaxial and abaxial view　D. Leaf, transversely sectioned　E. Staminate flower, lateral view　F. Staminate flower, abaxial view　G. Tepals of staminate flower, adaxial view　H. Tepals of staminate flower, abaxial view　I. Pistillate flower, lateral view　J. Pistillate flower, adaxial view　K. Pistillate flower, abaxial view　L. Tepals of pistillate flower, adaxial view　M. Tepals of pistillate flower, abaxial view　N. Capsule　O. Ovary, transversely sectioned　P. Gynoecium, adaxial view　Q. Styles and stigmas　R. Androecium, lateral view

扁果组 *Begonia* Sect. *Platycentrum* (Klotzsch) A. DC.

九九峰秋海棠

Begonia bouffordii C. I Peng

多年生草本，具根状茎，高约25 cm，地上茎无或非常短。叶基生或偶有茎生，叶柄紫红色；叶片斜卵形，肉质，近无毛，背面叶脉紫红色。花呈粉红色至白色；蒴果三棱形，具不等长3翅，背翅宽椭圆形或圆形，侧翅较小；子房2室，中轴胎座；花柱2。

分布： 原产于中国（台湾）。

生境： 生于海拔300～400 m干燥半荫蔽至荫蔽陡峭的被严重侵蚀的砾石上。

Perennial herb, rhizomatous, up to *ca.* 25 cm tall, stems absent or relatively short. Leaves basal or sometimes cauline, petioles reddish purple; blades oblique ovate, succulent, nearly glabrous, abaxial surface reddish purple on veins. Flowers pinkish to white; capsule trigonous, with 3 unequal wings, abaxial wing broadly elliptic to orbicular, lateral wings smaller; ovary 2-locular, placentae axile; styles 2.

Distribution: Native to China (Taiwan).

Habitat: On dry, semi-shaded to shaded, heavily eroded gravel slopes in steep ravines, 300-400 m elevation.

A

5cm

A. 植株　B. 花序　C. 叶片正反面　D. 叶片横切面　E. 雄花正面观　F. 雄花反面观　G. 雄花花被片正面观　H. 雄花花被片反面观　I. 雌花正面观　J. 雌花侧面观　K. 雌花花被片正面观　L. 雌花花被片反面观　M. 雄蕊　N. 蒴果　O. 子房横切面　P. 雌蕊　Q. 花柱和柱头

A. Habit　B. Inflorescence　C. Leaf, adaxial and abaxial view　D. Leaf, transversely sectioned　E. Staminate flower, adaxial view　F. Staminate flower, abaxial view　G. Tepals of staminate flower, adaxial view　H. Tepals of staminate flower, abaxial view　I. Pistillate flower, lateral view　J. Pistillate flower, adaxial view　K. Tepals of Pistillate flower, adaxial view　L. Tepals of pistillate flower, abaxial view　M. Androecium, lateral view　N. Capsule　O. Ovary, transversely sectioned　P. Gynoecium, lateral view　Q. Styles and stigmas

伯基尔秋海棠
Begonia burkillii Dunn

多年生草本，雌雄异株，具根状茎。叶片披针状卵形或卵形；表面蓝绿色，通常有交替的放射状明暗带或白色斑点；背面浅绿色，点缀不同深浅的猩红色。雌花白色至浅粉色；蒴果菱形，表面具微小腺毛，具近等长4翅；子房4室，中轴胎座；花柱4。

分布： 原产于中国、印度、缅甸。

生境： 常见于海拔213 ~ 1 188 m的溪边、阴凉处潮湿的岩石上。

Perennial herb, dioecious, rhizomatous. Leaf blades lanceolate-ovate to ovate; abaxial surface bluish-green, usually with alternate radiating light and dark bands or white blotches, adaxial surface pale green and shades of scarlet. Flowers white to pinkish, capsule rhomboid, with microscopic glandular hairs, subequal 4 wings; ovary 4-locular, placentae axile; styles 4.

Distribution: Native to China, India, and Myanmar.

Habitat: Locally common by streams, on wet rocks in deep shade, 213-1,188 m elevation.

A

10cm

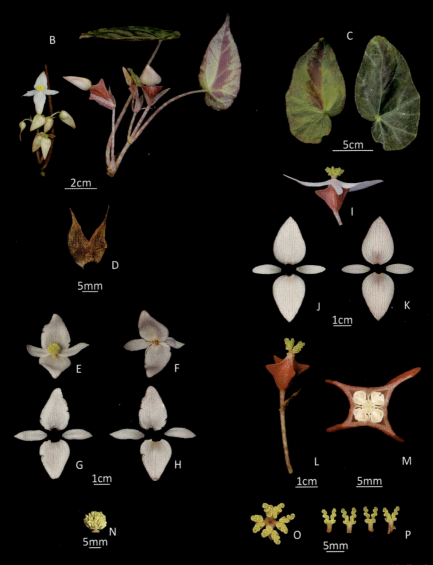

A. 植株　B. 花序　C. 叶片正反面　D. 托叶　E. 雄花正面观　F. 雄花反面观
G. 雄花花被片正面观　H. 雄花花被片反面观　I. 雌花侧面观　J. 雌花花被片正面观
K. 雌花花被片反面观　L. 蒴果　M. 子房横切面　N. 雄蕊　O. 雌蕊　P. 花柱和柱头

A. Habit　B. Inflorescence　C. Leaf, adaxial and abaxial view　D. Stipule, adaxial view
E. Staminate flower, adaxial view　F. Staminate flower, abaxial view　G. Tepals of staminate
flower, adaxial view　H. Tepals of staminate flower, abaxial view　I. Pistillate flower, lateral
view　J. Tepals of Pistillate flower, adaxial view　K. Tepals of pistillate flower, abaxial view
L. Capsules　M. Ovary, transversely sectioned　N. Androecium, lateral view　O. Gynoecium,
adaxial view　P. Styles and stigmas

花叶秋海棠

Begonia cathayana Hemsl.

多年生草本，地上茎直立，常分枝，被褐色柔毛。叶茎生，叶片卵形至宽卵形，薄纸质；表面深绿色，具V形的紫红色环纹，背面淡绿色或紫红色，掌状叶脉明显。花白色，粉红色或橘黄色；蒴果倒卵状长圆形，密被毛，具不等3翅；子房2室，中轴胎座；花柱2。

分布： 原产于中国（广西、云南、广东）、越南。

生境： 混交林下阴凉、潮湿处。

Perennial herb, stems erect, often branched, red puberulous. Leaves cauline, blades ovate to broadly ovate, thin papery; adaxial surface dark green with a purple-red V-shaped band, abaxial surface light green or purple-red, palmate veins prominent. Flowers white, pink or, orangish; capsule obovate oblong, densely hairy, with unequal 3 wings; ovary 2-locular, placentae axile; styles 2.

Distribution: Native to China (Guangxi, Yunnan, Guangdong), and Vietnam.

Habitat: Scrubby vegetation or forests, in shaded environments.

A

10cm

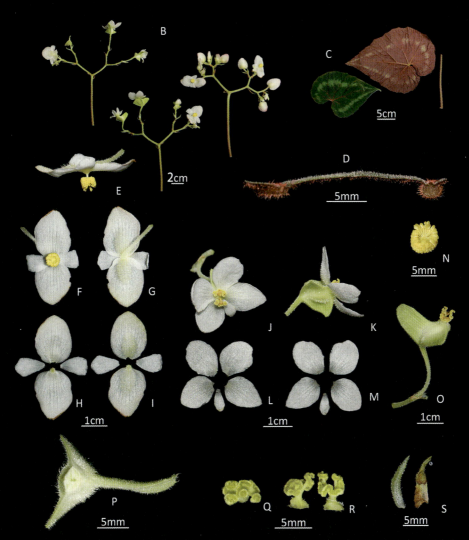

A. 植株　B. 花序　C. 叶片正反面和叶柄　D. 叶片横切面　E. 雄花侧面观　F. 雄花正面观　G. 雄花反面观　H. 雄花花被片正面观　I. 雄花花被片反面观　J. 雌花正面观　K. 雌花侧面观　L. 雌花花被片正面观　M. 雌花花被片反面观　N. 雄蕊　O. 蒴果　P. 子房横切面　Q. 雌蕊　R. 花柱和柱头　S. 苞片

A. Habit　B. Inflorescence　C. Leaf, adaxial and abaxial view, and petiole　D. Leaf, transversely sectioned　E. Staminate flower, lateral view　F. Staminate flower, adaxial view　G. Staminate flower, abaxial view　H. Tepals of staminate flower, adaxial view　I. Tepals of staminate flower, abaxial view　J. Pistillate flower, adaxial view　K. Pistillate flower, lateral view　L. Tepals of pistillate flower, adaxial view　M. Tepals of pistillate flower, abaxial view　N. Androecium, adaxial view　O. Capsule　P. Ovary, transversely sectioned　Q. Gynoecium, adaxial view　R. Styles and stigmas　S. Bracts

赤水秋海棠

Begonia chishuiensis T. C. Ku

多年生草本，根状茎棕色。叶基生，叶片长卵形至宽披针形，边缘有浅齿或呈缺刻状；表面褐绿色，具短硬毛，背面淡褐绿色，近无毛，沿脉被长硬毛。花白色或粉色；蒴果长卵形，无毛或近无毛，具不等长3翅，背翅大，呈宽镰刀状，侧翅较小，呈半月形；子房2室，中轴胎座；花柱2。

分布：原产于中国（贵州）。

生境：潮湿的岩石上。

Perennial herb, rhizomes brown. Leaves basal, blades long ovate to broadly lanceolate, margin shallowly toothed or notched; adaxial surface brown-green, with short, stiff hairs, abaxial surface light brown-green, subglabrous, long stiff hairs along veins. Flowers white or pinkish; capsule long ovate, glabrous or subglabrous, with unequal 3 wings, abaxial wing broadly falcate, lateral wings smaller, lunate; ovary 2-locular, placentae axile; styles 2.

Distribution: Native to China (Guizhou).

Habitat: On moist rocks.

A

5cm

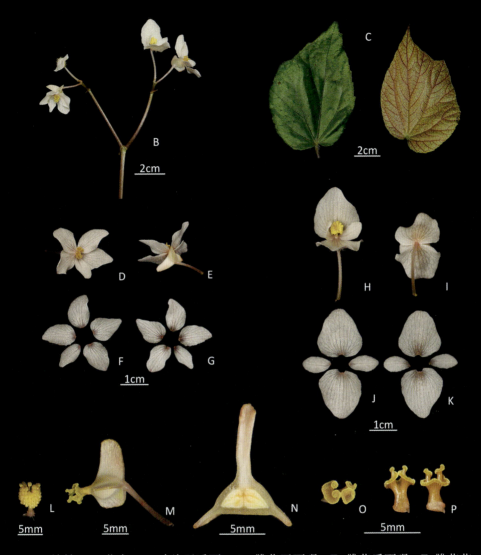

A. 植株　B. 花序　C. 叶片正反面　　D. 雌花正面观　E. 雌花反面观　F. 雌花花被片正面观　G. 雌花花被片反面观　H. 雄花正面观　I. 雄花反面观　J. 雄花花被片正面观　K. 雄花花被片反面观　L. 雄蕊　M. 蒴果　N. 子房横切面　O. 雌蕊　P. 花柱和柱头

A. Habit　B. Inflorescence　C. Leaf, adaxial and abaxial view　D. Pistillate flower, adaxial view　E. Pistillate flower, abaxial view　F. Tepals of pistillate flower, adaxial view G. Tepals of pistillate flower, abaxial view　H. Staminate flower, adaxial view　I. Staminate flower, abaxial view　J. Tepals of staminate flower, adaxial view　K. Tepals of staminate flower, abaxial view　L. Androecium, lateral view　M. Capsule　N. Ovary, transversely sectioned　O. Gynoecium, adaxial view　P. Styles and stigmas

溪头秋海棠

Begonia chitoensis Tang S. Liu & M. J. Lai

多年生草本，具根状茎，高40～95 cm。叶基生和茎生，叶片宽卵形到圆形；表面近无毛或稀被毛，背面近无毛。花序腋生；蒴果下垂，椭圆形，无毛，具不等长3翅；子房2室，中轴胎座；花柱2。

分布： 原产于中国（台湾）。

生境： 海拔400～2 200 m阔叶林下的遮蔽潮湿处。

Perennial herb, rhizomatous, 40-95 cm tall. Leaves basal and cauline, blades broadly ovate to orbicular; abaxial surface subglabrous to sparsely hairy, adaxial surface subglabrous. Inflorescences axillary; capsule nodding, ellipsoid, glabrous, with unequal 3 wings; ovary 2-locular, placentae axile; styles 2.

Distribution: Native to China (Taiwan).

Habitat: Shaded, moist, broad-leaved forests; 400-2,200 m.

A

5cm

A. 植株　B. 花序　C. 叶片正反面　D. 雄花正面观　E. 雄花反面观　F. 雄花花被片正面观　G. 雄花花被片反面观　H. 雌花正面观　I. 雌花反面观　J. 雌花花被片正面观　K. 雌花花被片反面观　L. 雄蕊　M. 蒴果　N. 子房横切面　O. 雌蕊　P. 花柱和柱头

A. Habit　B. Inflorescence　C. Leaf, adaxial and abaxial view　D. Staminate flower, adaxial view　E. Staminate flower, abaxial view　F. Tepals of staminate flower, adaxial view　G. Tepals of staminate flower, abaxial view　H. Pistillate flower, adaxial view　I. Pistillate flower, abaxial view　J. Tepals of pistillate flower, adaxial view　K. Tepals of pistillate flower, abaxial view　L. Androecium, lateral view　M. Capsule　N. Ovary, transversely sectioned　O. Gynoecium, adaxial view　P. Styles and stigmas

厚叶秋海棠

Begonia dryadis Irmsch.

多年生草本，具根状茎，地上茎直立。叶片卵圆形至宽卵形，近革质，褶皱，无毛；表面深绿色，背面浅绿色。花粉色，雄花花被片4，雌花花被片5；蒴果呈三棱椭圆形，红色，具不等长3翅；子房2室，中轴胎座；花柱2。

分布： 原产于中国（云南）。

生境： 林下沟谷、溪边阴湿处。

Perennial herb, rhizomatous, stems erect. Leaf blades ovate-orbicular to broadly ovate, subleathery, rugulose, glabrous; adaxial surface dark green, abaxial surface light green. Flowers pinkish; male flowers tepals 4, female flowers tepals 5; capsule trigonous-ellipsoid, red, with unequal 3 wings; ovary 2-locular, placentae axile; styles 2.

Distribution: Native to China (Yunnan).

Habitat: In shaded, moist valleys, or along streamsides in forests.

A

5cm

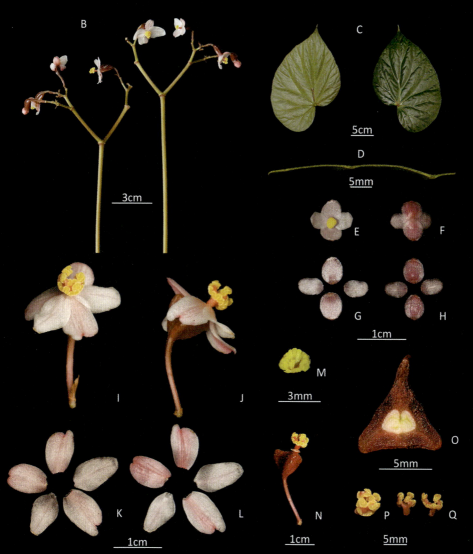

A. 植株　B. 花序　C. 叶片正反面　D. 叶片横切面　E. 雄花正面观　F. 雄花反面观 G. 雄花花被片正面观　H. 雄花花被片反面观　I. 雌花正面观　J. 雌花侧面观　K. 雌 花花被片正面观　L. 雌花花被片反面观　M. 雄蕊　N. 蒴果　O. 子房横切面　P. 雌 蕊　Q. 花柱和柱头

A. Habit　B. Inflorescence　C. Leaf, adaxial and abaxial view　D. Leaf, transversely sectioned　E. Staminate flower, adaxial view　F. Staminate flower, abaxial view　G. Tepals of staminate flower, adaxial view　H. Tepals of staminate flower, abaxial view　I. Pistillate flower, adaxial view　J. Pistillate flower, lateral view　K. Tepals of pistillate flower, adaxial view　L. Tepals of pistillate flower, abaxial view　M. Androecium, adaxial view　N. Capsule O. Ovary, transversely sectioned　P. Gynoecium, adaxial view　Q. Styles and stigmas

川边秋海棠

Begonia duclouxii Gagnep.

多年生草本，根状茎伸长。叶基生，叶柄被锈褐色卷曲长毛；叶片卵形，不对称，基部深心形，边缘常呈不规则的锯齿状；表面被毛，叶背面沿脉被长柔毛。二歧聚伞花序，光滑无毛；花红色至粉红色；蒴果被柔毛，具不等3翅；子房2室，中轴胎座；花柱2。

分布： 原产于中国（云南）。

生境： 海拔1 000 ~ 1 400 m。

Perennial herb, rhizomes elongate. Leaves basal, petioles rusty-brown curled villous; blades ovate, asymmetric, base deeply cordate, margin irregularly serrate, adaxial surface hairy, abaxial surface villous along venation. Inflorescences dichocyme, glabrous; flowers red or pinkish; capsule villous, with unequal 3 wings; ovary 2-locular, placentae axile; styles 2.

Distribution: Native to China (Yunnan).

Habitat: 1,000-1,400 m elevation.

A

5cm

A. 植株　B. 花序　C. 叶片正反面　D. 雄花正面观　E. 雄花反面观　F. 雄花花被片正面观　G. 雄花花被片反面观　H. 雌花正面观　I. 雌花侧面观　J. 雌花花被片正面观　K. 雌花花被片反面观　L. 雄蕊　M. 蒴果　N. 子房横切面　O. 雌蕊　P. 花柱和柱头

A. Habit　B. Inflorescence　C. Leaf, adaxial and abaxial view　D. Staminate flower, adaxial view　E. Staminate flower, abaxial view　F. Tepals of staminate flower, adaxial view　G. Tepals of staminate flower, abaxial view　H. Pistillate flower, adaxial view　I. Pistillate flower, lateral view　J. Tepals of pistillate flower, adaxial view　K. Tepals of pistillate flower, abaxial view　L. Androecium, adaxial view　M. Capsule　N. Ovary, transversely sectioned O. Gynoecium, adaxial view　P. Styles and stigmas

食用秋海棠

Begonia edulis H. Lév.

多年生草本，具根状茎伸长，地上茎直立，有沟纹和疣点，无毛或近无毛。叶片圆形或扁球形，稍不对称，浅裂至叶长1/4处。花粉色，无毛；蒴果无毛，具不等长3翅；子房2室，中轴胎座；花柱2。

分布： 原产于中国、越南。

生境： 林下或潮湿的岩石上。

Perennial herb, rhizomes elongate, stems erect, with furrows and warts, glabrous or subglabrous. Leaf blades orbicular or oblate-orbicular, slightly asymmetric, and shallowly lobed to 1/4 of leaf length. Flowers pinkish, glabrous; capsule glabrous, with unequal 3 wings; 2-locular, placentae axile; styles 2.

Distribution: Native to China and Vietnam.

Habitat: On the forest floor or shaded moist rocks.

A

10cm

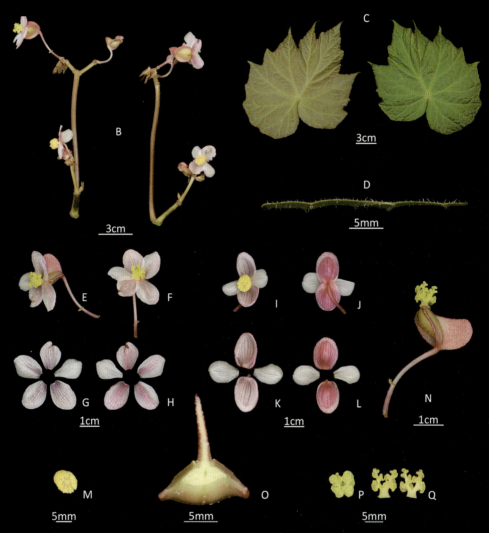

A. 植株　B. 花序　C. 叶片正反面　D. 叶片横切面　E. 雌花侧面观　F. 雌花正面观 G. 雌花花被片正面观　H. 雌花花被片反面观　I. 雄花正面观　J. 雄花反面观　K. 雄花花被片正面观　L. 雄花花被片反面观　M. 雄蕊　N. 蒴果　O. 子房横切面　P. 雌蕊　Q. 花柱和柱头

A. Habit　B. Inflorescence　C. Leaf, adaxial and abaxial view　D. Leaf, transversely sectioned　E. Pistillate flower, lateral view　F. Pistillate flower, adaxial view　G. Tepals of pistillate flower, adaxial view　H. Tepals of pistillate flower, abaxial view　I. Staminate flower, adaxial view　J. Staminate flower, abaxial view　K. Tepals of staminate flower, adaxial view　L. Tepals of staminate flower, abaxial view　M. Androecium, adaxial view N. Capsule　O. Ovary, transversely sectioned　P. Gynoecium, adaxial view　Q. Styles and stigmas

水鸭脚

Begonia formosana (Hayata) Masam.

多年生草本，根状茎伸长，地上茎直立，高35～95 cm，无毛或近无毛。叶基部和茎生；叶片卵形至宽卵形，不对称，形似水鸭的脚。花序近无毛；花白色至粉红色；蒴果无毛，具不等长3翅；子房2室，中轴胎座；花柱2。

分布：原产于中国（台湾）。

生境：阴坡林下潮湿地，海拔700～900 m。

Perennial herb, rhizomes elongate, stems erect, 35-95 cm tall, glabrous or subglabrous. Leaves basal and cauline; blades ovate to broadly ovate, asymmetric, resembling duck feet. Inflorescences subglabrous; flowers white to pinkish; capsule glabrous, with unequal 3 wings; ovary 2-locular, placentae axile; styles 2.

Distribution: Native to China (Taiwan).

Habitat: Forests, shaded, moist environments, 700-900 m elevation.

10cm

A

10cm

　　A. 植株　B. 茎，腋生花序　C. 花序　D. 叶片正反面　E. 叶片横切面　F. 雄花正面观　G. 雄花反面观　H. 雄花花被片正面观　I. 雄花花被片反面观　J. 雌花反面观　K. 雌花正面观　L. 雌花花被片反面观　M. 雌花花被片正面观　N. 雄蕊　O. 雌蕊　P. 花柱和柱头　Q. 子房横切面　R. 蒴果

　　A. Habit　B. Stem, inflorescence axillary　C. Inflorescence　D. Leaf, adaxial and abaxial view　E. Leaf, transversely sectioned　F. Staminate flower, adaxial view　G. Staminate flower, abaxial view　H. Tepals of staminate flower, adaxial view　I. Tepals of staminate flower, abaxial view　J. Pistillate flower, abaxial view　K. Pistillate flower, adaxial view　L. Tepals of pistillate flower, abaxial view　M. Tepals of pistillate flower, adaxial view　N. Androecium, adaxial view　O. Gynoecium, adaxial view　P. Styles and stigmas　Q. Ovary, transversely sectioned　R. Capsule

铺地秋海棠

Begonia handelii var. **prostrata** (Irmsch.) Tebbitt

多年生草本，雌雄异株，具根茎。叶片卵状长圆形，无毛，叶脉红褐色。花序通常生于茎基部，花白色至粉色；蒴果表面具疣点，无翅或4～8个翅；子房4室，中轴胎座；花柱4。

分布： 原产于中国、越南、老挝、缅甸、泰国。

生境： 林下坡地。

Perennial herb, dioecious, rhizomatous. Leaf blades ovate-oblong, glabrous, veins reddish-brown. Inflorescences typically borne at the stem base, flowers white to pinkish; capsule with warts, wingless or 4-8 wings; ovary 4-locular, placentae axile; styles 4.

Distribution: Native to China, Vietnam, Laos, Myanmar, and Thailand.

Habitat: Forests, on mountain slopes.

A

10cm

A. 植株　B. 花序　C. 叶片正反面　D. 雄花正面观　E. 雄花反面观　F. 雄花花被片正面观　G. 雄花花被片反面观　H. 雌花正面观　I. 雌花反面观　J. 雌花花被片正面观　K. 雌花花被片反面观　L. 雄蕊　M. 蒴果　N. 子房横切面　O. 雌蕊　P. 花柱和柱头　Q. 苞片

A. Habit　B. Inflorescence　C. Leaf, adaxial and abaxial view　D. Staminate flower, adaxial view　E. Staminate flower, abaxial view　F. Tepals of staminate flower, adaxial view　G. Tepals of staminate flower, abaxial view　H. Pistillate flower, adaxial view　I. Pistillate flower, abaxial view　J. Tepals of pistillate flower, adaxial view　K. Tepals of pistillate flower, abaxial view　L. Androecium, adaxial view　M. Capsule　N. Ovary, transversely sectioned　O. Gynoecium, adaxial view　P. Styles and stigmas　Q. Bracts, adaxial and abaxial view

长翅秋海棠

Begonia longialata K. Y. Guan & D. K. Tian

多年生草本，全株无毛。叶片基生和茎生，叶柄具红色线型斑纹；叶片近圆形，稍不对称，基部深心形，边缘有不规则锯齿，掌状深裂到叶长的2/3。花序光滑无毛，花白色至粉色；蒴果具极不等长3翅，背翅硕大；子房2室，中轴胎座；花柱2。

分布： 原产于中国（云南）。

生境： 潮湿的岩石斜坡上。

Perennial herb, glabrous throughout. Leaves basal and cauline, petioles with many red linear dots; blades suborbicular, slightly asymmetric, base deeply cordate, margin irregularly serrate, distinctly divided to 2/3 of leaf length. Inflorescences glabrous, flowers white to pinkish; capsule with highly unequal 3 wings, abaxial wing much protruded; ovary 2-locular, placentae axile; styles 2.

Distribution: Native to China (Yunnan).

Habitat: On moist rocky slopes.

A

10cm

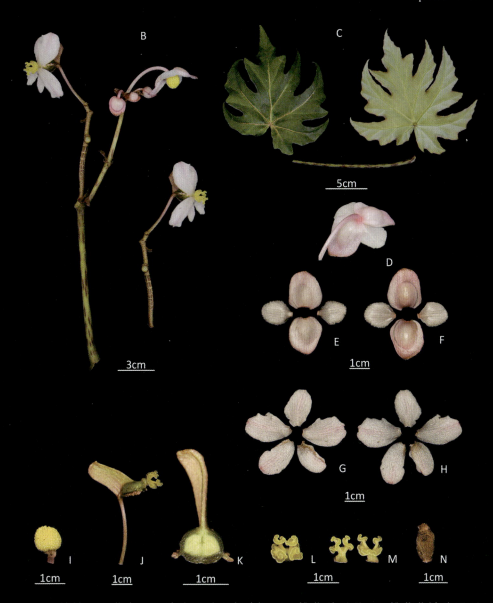

A. 植株　B. 花序　C. 叶片正反面和叶柄　D. 雄花反面观　E. 雄花花被片正面观　F. 雄花花被片反面观　G. 雌花花被片正面观　H. 雌花花被片反面观　I. 雄蕊　J. 蒴果　K. 子房横切面　L. 雌蕊　M. 花柱和柱头　N. 苞片

A. Habit　B. Inflorescence　C. Leaf, adaxial and abaxial view, and petiole　D. Staminate flower, abaxial view　E. Tepals of staminate flower, adaxial view　F. Tepals of staminate flower, abaxial view　G. Tepals of pistillate flower, adaxial view　H. Tepals of pistillate flower, abaxial view　I. Androecium, lateral view　J. Capsule　K. Ovary, transversely sectioned L. Gynoecium, adaxial view　M. Styles and stigmas　N. Bract

长纤秋海棠

Begonia longiciliata C. Y. Wu

多年生草本，具根茎。叶片卵形或近圆形，边缘被长毛；表面具有白色环斑，疏被长硬毛；叶柄棕红色，多毛。花白色至粉红色；雄花花被片4，雌花花被片5；蒴果光滑无毛，具不等长3翅，背翅硕大；子房2室，中轴胎座；花柱3。

分布：原产于中国（云南、贵州、广西）、越南、老挝。

生境：山沟岩石斜坡上和山沟密林中。

Perennial herb, rhizomatous. Leaves ovate to orbicular, margin covered with long hairs; adaxial surface with white circular markings, sparsely hirsute; petioles brownish red, hirsute-villous. Flowers white to pinkish; male flowers tepals 4, female flowers tepals 5; capsule glabrous, with unequal 3 wings, abaxial wing much protruded; ovary 2-locular, placentae axile; styles 2.

Distribution: Native to China (Yunnan, Guizhou, Guangxi), Vietnam, and Laos.

Habitat: On rocky slopes and in dense forests of mountain valleys.

A

10cm

A. 植株　B. 花序　C. 叶片正反面　D. 叶片横切面　E. 雄花正面观　F. 雄花反面观　G. 雄花花被片正面观　H. 雄花花被片反面观　I. 雌花正面观　J. 雌花反面观　K. 雌花花被片正面观　L. 雌花花被片反面观　M, N. 雄蕊　O. 蒴果　P. 子房横切面　Q. 雌蕊　R. 花柱和柱头

A. Habit　B. Inflorescence　C. Leaf, adaxial and abaxial view　D. Leaf, transversely sectioned　E. Staminate flower, adaxial view　F. Staminate flower, abaxial view　G. Tepals of staminate flower, adaxial view　H. Tepals of staminate flower, abaxial view　I. Pistillate flower, adaxial view. J. Pistillate flower, abaxial view　K. Tepals of pistillate flower, adaxial view　L. Tepals of pistillate flower, abaxial view　M, N. Androecium, lateral and adaxial view　O. Capsule　P. Ovary, transversely sectioned　Q. Gynoecium, adaxial view　R. Styles and stigmas

山地秋海棠

Begonia oreodoxa Chun & F. Chun

多年生草本，根状茎伸长。叶基生，叶柄密被红褐色柔毛；叶片卵形至近圆形，表面绿色，叶片基部有较暗的区域。花序光滑无毛；花白色至粉色；蒴果通常被长柔毛，偶无毛，具不等长3翅，翅小；子房2室，中轴胎座；花柱2。

分布：原产于中国（云南）、越南北部。

生境：荫蔽潮湿的环境，溪边山坡上。

Perennial herb, rhizomes elongate. Leaves basal, petioles densely reddish brown villous; blades ovate-suborbicular, adaxial surface green with dark areas particularly at base. Inflorescences glabrous; flowers white to pinkish; capsule villous or rarely glabrous, with unequal 3 wings, wings small; ovary 2-locular, placentae axile; styles 2.

Distribution: Native to China (Yunnan) and Northern Vietnam.

Habitat: Shaded moist environments on mountain slopes or by streams.

A

5cm

A. 植株　B. 花序　C. 叶片正反面　D. 雄花侧面观　E. 雄花反面观　F. 雄花花被片正面观　G. 雄花花被片反面观　H. 雌花正面观　I. 雌花反面观　J. 雌花花被片正面观 K. 雌花花被片反面观　L. 雄蕊　M. 蒴果　N. 子房横切面　O. 雌蕊　P. 花柱和柱头

A. Habit　B. Inflorescence　C. Leaf, adaxial and abaxial view　D. Staminate flower, lateral view　E. Staminate flower, abaxial view　F. Tepals of staminate flower, adaxial view G. Tepals of staminate flower, abaxial view　H. Pistillate flower, adaxial view　I. Pistillate flower, abaxial view　J. Tepals of pistillate flower, adaxial view　K. Tepals of pistillate flower, abaxial view　L. Androecium, lateral view　M. Capsule　N. Ovary, transversely sectioned　O. Gynoecium, adaxial view　P. Styles and stigmas

光滑秋海棠

Begonia psilophylla Irmsch.

多年生草本，根状茎块状或近球形，茎直立，仅花期可见，无毛。叶基生和茎生，互生；叶片卵形至心形，边缘有细锯齿，锯齿之间相隔较远，叶尖有尾巴状突出。花粉色至深红棕色；蒴果倒卵状长圆形，无毛，具不等长3翅；子房2室，中轴胎座；花柱2。

分布：原产于中国（云南）。

生境：林下石灰岩壁上，荫蔽潮湿的环境。

Perennial herb, rhizomes blocky or subglobose, stems erect, seen only at anthesis, glabrous. Leaves basal and cauline, alternate; blades ovate-cordate, margin remotely and minutely serrulate, apex caudate. Flowers pink to dark red-brown; capsule obovate oblong, glabrous, with unequal 3 wings; ovary 2-locular, placentae axile; styles 2.

Distribution: Native to China (Yunnan).

Habitat: Forests, on limestone rocks in shaded moist environments.

A

5cm

A. 植株　B. 花序　C. 叶片正反面　D. 叶片横切面　E. 雄花正面观　F. 雄花反面观　G. 雄花花被片正面观　H. 雄花花被片反面观　I. 雌花正面观　J. 雌花侧面观　K. 雌花花被片正面观　L. 雌花花被片反面观　M. 雄蕊　N. 蒴果　O. 子房横切面　P. 雌蕊　Q. 花柱和柱头

A. Habit　B. Inflorescence　C. Leaf, adaxial and abaxial view　D. Leaf, transversely sectioned　E. Staminate flower, adaxial view　F. Staminate flower, abaxial view　G. Tepals of staminate flower, adaxial view　H. Tepals of staminate flower, abaxial view　I. Pistillate flower, adaxial view　J. Pistillate flower, lateral view　K. Tepals of pistillate flower, adaxial view　L. Tepals of pistillate flower, abaxial view　M. Androecium, lateral view　N. Capsule O. Ovary, transversely sectioned　P. Gynoecium, adaxial view　Q. Styles and stigmas

大王秋海棠

Begonia rex Putz.

多年生草本，根状茎红褐色。叶基生，叶柄浅红色；叶片卵形至宽卵形，先端急尖至稍尖；表面深绿色，主脉间具淡白色斑块；背面紫绿色。花浅粉色，无毛；蒴果卵形，无毛，具不等长3翅，背翅硕大；子房2室，中轴胎座；花柱2。

分布： 原产于中国、孟加拉国、印度、缅甸和不丹。

生境： 森林、岩石上及山谷的洞里。

Perennial herb, rhizomes reddish brown. Leaves basal, petioles light red; blades ovate to broadly ovate, apex acute to shortly acuminate; adaxial surface dark green, pale white patches between main veins, abaxial surface purplish-green. Flowers pinkish, glabrous; capsule ovate, glabrous, with unequal 3 wings, abaxial wing much protruded; ovary 2-locular, placentae axile; styles 2.

Distribution: Native to China, Bangladesh, India, Myanmar, and Bhutan.

Habitat: Forests, on rocks, and in caves in valleys.

A

10cm

A. 植株　B. 花序　C. 叶片正反面　D. 叶片横切面　E. 雌花侧面观　F. 雌花正面观　G. 雌花反面观　H. 雌花花被片正面观　I. 雌花花被片反面观　J. 雄花正面观　K. 雄花反面观　L. 雄花花被片正面观　M. 雄花花被片反面观　N. 雄蕊　O. 蒴果　P. 子房横切面　Q. 雌蕊　R. 花柱和柱头

A. Habit　B. Inflorescence　C. Leaf, adaxial and abaxial view　D. Leaf, transversely sectioned　E. Pistillate flower, lateral view　F. Pistillate flower, adaxial view　G. Pistillate flower, abaxial view　H. Tepals of pistillate flower, adaxial view　I. Tepals of pistillate flower, abaxial view　J. Staminate flower, adaxial view　K. Staminate flower, adaxial view　L. Tepals of staminate flower, adaxial view　M. Tepals of staminate flower, abaxial view　N. Androecium, lateral view　O. Capsule　P. Ovary, transversely sectioned　Q. Gynoecium, adaxial view　R. Styles and stigmas

深圳秋海棠

Begonia shenzhenensis D. K. Tian & X. Yun Wang

多年生草本，具根状茎。单叶互生，基生，冬季落叶；叶柄淡绿色至粉红色；叶片斜卵圆形至心形，叶缘有锯齿，先端渐尖或短尾状；表面绿色，粗糙，具短的浅灰色硬毛。花白色，子房附近宿存一对淡绿色小苞片；蒴果绿色，具不等长3翅；子房2室，中轴胎座；花柱2。

分布： 原产于中国（广东）。

生境： 林冠下小溪旁岩石上。

Perennial herb, rhizomatous. Leaves simple, basal, alternate, deciduous in winter; petioles light green to pinkish; blades obliquely ovate-cordate, margins serrate, apex acuminate or short-caudate; adaxial surface green, rough, with short grayish strigose. Flowers white, a pair of pale green bracteoles persistent near ovary; capsule greenish, with unequal 3 wings; ovary 2-locular, placentae axile; styles 2.

Distribution: Native to China (Guangdong).

Habitat: On rocks along a small stream beneath the forest canopy.

A

5cm

A. 植株　B. 花序　C. 叶片正反面　D. 叶片横切面　E. 雄花正面观　F. 雄花反面观　G. 雄花花被片正面观　H. 雄花花被片反面观　I. 雄蕊　J. 子房横切面

A. Habit　B. Inflorescence　C. Leaf, adaxial and abaxial view　D. Leaf, transversely sectioned　E. Staminate flower, adaxial view　F. Staminate flower, abaxial view　G. Tepals of staminate flower, adaxial view　H. Tepals of staminate flower, abaxial view　I. Androecium, adaxial view　J. Ovary, transversely sectioned

高茎秋海棠

Begonia sphenantheroides C. I Peng

多年生草本，茎直立，株高可达150 cm。叶片卵形至宽卵形，无毛或近无毛，边缘具不规则松散细锯齿；表面绿色，背面红色至淡红色。花鲜红色；蒴果红色，具不等长3翅，翅红色；子房2室，中轴胎座；花柱2。

分布： 原产于越南东北部。

生境： 溪流上方岩石斜坡，荫蔽潮湿的环境。

Perennial herb, stems erect, up to *ca*. 150 cm tall. Leaf blades ovate to broadly ovate, glabrous or subglabrous, margin irregularly loosely serrulate; adaxial surface green, abaxial surface reddish to pale red. Flowers bright red; capsule red, with unequal 3 wings, wings red; ovary 2-locular, placentae axile; styles 2.

Distribution: Native to Northeastern Vietnam.

Habitat: On rocky slopes above a running stream in a shaded and moist environment.

A

10cm

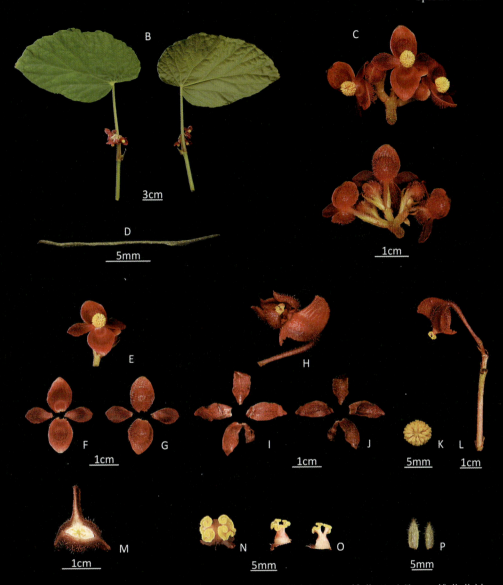

A. 植株　B. 叶片正反面　C. 花序　D. 叶片横切面　E. 雄花正面观　F. 雄花花被片正面观　G. 雄花花被片反面观　H. 雌花侧面观　I. 雌花花被片正面观　J. 雌花花被片反面观　K. 雄蕊　L. 蒴果　M. 子房横切面　N. 雌蕊　O. 花柱和柱头　P. 苞片

A. Habit　B. Leaf, adaxial and abaxial view　C. Inflorescence　D. Leaf, transversely sectioned　E. Staminate flower, adaxial view　F. Tepals of staminate flower, adaxial view G. Tepals of staminate flower, abaxial view　H. Pistillate flower, lateral view　I. Tepals of pistillate flower, adaxial view　J. Tepals of pistillate flower, abaxial view　K. Androecium, adaxial view　L. Capsule　M. Ovary, transversely sectioned　N. Gynoecium, adaxial view O. Styles and stigmas　P. Bracts, adaxial and abaxial view

汤姆森秋海棠

Begonia thomsonii A. DC.

多年生草本，根状茎强壮。叶互生；叶柄圆柱状，淡绿色至红色，密被白色短绒毛；叶片宽卵形至圆形，全缘；表面橙绿色，具长柔毛。总状花序腋生；花梗绿色，具长柔毛；蒴果三棱椭圆形，绿色，具不等长3翅；子房2室，中轴胎座；花柱2。

分布： 原产于中国、印度、孟加拉国和缅甸。

生境： 潮湿的林下环境。

Perennial herb, rhizomes stout. Leaves alternate; petioles terete, pale green to reddish, densely white velutinous; blades broadly ovate to orbicular, margin entire; adaxial surface lime green, villous. Inflorescences axillary, cymose; peduncle greenish, with villous; capsule trigonous-ellipsoid, greenish, with unequal 3 wings; ovary 2-locular, placentae axile; styles 2.

Distribution: Native to China, India, Bangladesh, and Myanmar.

Habitat: In moist forests.

A

5cm

A. 植株　B. 花序　C. 叶片正反面　D. 叶片横切面　E. 雄花正面观　F. 雄花反面观
G. 雄花花被片正面观　H. 雄花花被片反面观　I, J. 雌花侧面观　K. 雌花花被片正面观
L. 雌花花被片反面观　M. 雄蕊　N. 蒴果　O. 子房横切面　P. 雌蕊　Q. 花柱和柱头

A. Habit　B. Inflorescence　C. Leaf, adaxial and abaxial view　D. Leaf, transversely sectioned　E. Staminate flower, adaxial view　F. Staminate flower, abaxial view　G. Tepals of staminate flower, adaxial view　H. Tepals of staminate flower, abaxial view　I, J. Pistillate flower, lateral view　K. Tepals of pistillate flower, adaxial view　L. Tepals of pistillate flower, abaxial view　M. Androecium, adaxial view　N. Capsule　O. Ovary, transversely sectioned　P. Gynoecium, adaxial view　Q. Styles and stigmas

钟扬秋海棠

Begonia zhongyangiana W. G. Wang & S. Z. Zhang

　　多年生草本，根状茎短，很少分枝。叶片宽卵形，绿色，叶脉间具白色条带，两面密被长柔毛。花橙黄色，雄花先开放；蒴果密被白色长柔毛，具不等长3翅；子房2室，中轴胎座；花柱2。

　　分布：原产于中国（西藏）。

　　生境：海拔693 ~ 1 600 m的常绿阔叶山地森林下的岩石或斜坡上。

Perennial herb, rhizomes short, rarely branched. Leaf blades broadly ovate, green, variegated with interveinal white stripes, both surfaces densely white villous. Flowers orange-yellow, staminate flower opening earlier than pistillate flower; capsule densely white villous, with unequal 3 wings; ovary 2-locular, placentae axile; styles 2.

　　Distribution: Native to China (Xizang).

　　Habitat: On rocks or slopes under the ridges in broadleaved evergreen montane forest, 693-1,600 m elevation.

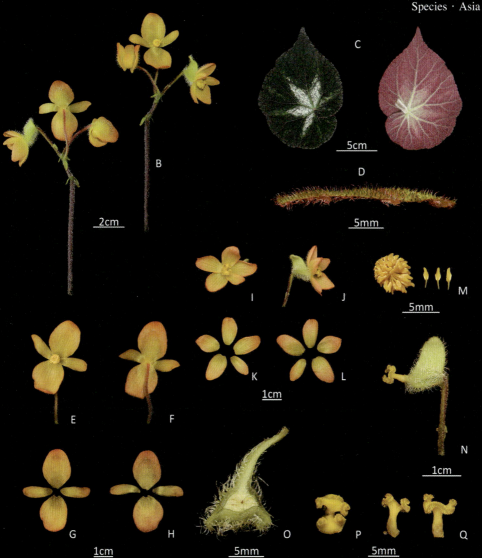

A. 植株　B. 花序　C. 叶片正反面　D. 叶片横切面　E. 雄花正面观　F. 雄花反面观 G. 雄花花被片正面观　H. 雄花花被片反面观　I. 雌花正面观　J. 雌花侧面观　K. 雌花花被片正面观　L. 雌花花被片反面观　M. 雄蕊　N. 蒴果　O. 子房横切面　P. 雌蕊　Q. 花柱和柱头

A. Habit　B. Inflorescence　C. Leaf, adaxial and abaxial view　D. Leaf, transversely sectioned　E. Staminate flower, adaxial view　F. Staminate flower, abaxial view　G. Tepals of staminate flower, adaxial view　H. Tepals of staminate flower, abaxial view　I. Pistillate flower, adaxial view　J. Pistillate flower, lateral view　K. Tepals of pistillate flower, adaxial view　L. Tepals of pistillate flower, abaxial view　M. Androecium, adaxial view　N. Capsule O. Ovary, transversely sectioned　P. Gynoecium, adaxial view　Q. Styles and stigmas

侧膜组 *Begonia* Sect. *Coelocentrum* Irmsch.

蛛网脉秋海棠

Begonia arachnoidea C. I Peng, Yan Liu & S. M. Ku

多年草本，根状茎强壮，匍匐。叶基生，盾状；叶片近圆形或宽卵形，纸质；表面深绿色或略带褐色，沿主脉有白色或浅色条纹，背面沿脉密被短刚毛；边缘具不均匀的浅细锯齿或波状；叶脉呈蜘蛛网状。花粉红色，蒴果长圆形，具不等长3翅；子房1室，侧膜胎座；花柱3。

分布： 原产于中国（广西）。

生境： 海拔200 m左右，半荫蔽竹林和灌木林的岩石坡上。

Perennial herb, rhizomes stout, creeping. Leaves basal, peltate; blades suborbicular or broadly ovate, papery; adaxial surface intensely green or brownish, with white or pale band along major veins, abaxial surface densely hispidulous-pilose on all veins; margin shallowly unequally serrulate or undulate; veins web-like. Flowers pink, capsule oblong, with unequal 3 wings; ovary 1-locular, placentation parietal; styles 3.

Distribution: Native to China (Guangxi).

Habitat: In mixed bamboo and scrubby vegetation at the foot of a limestone hill, on a semi-shaded rocky slope, *ca.* 200 m elevation.

A

5cm

A. 植株　B. 花序　C. 叶片正反面　D. 叶片横切面　E. 雄花正面观　F. 雄花反面观
G. 雄花花被片正面观　H. 雄花花被片反面观　I. 雌花侧面观　J. 雌花正面观　K. 雌花反面观　L. 雌花花被片正面观　M. 雌花花被片反面观　N. 蒴果　O. 子房横切面
P. 雌蕊　Q. 花柱和柱头　R. 雄蕊

A. Habit　B. Inflorescence　C. Leaf, adaxial and abaxial view　D. Leaf, transversely
sectioned　E. Staminate flower, adaxial view　F. Staminate flower, abaxial view　G. Tepals
of staminate flower, adaxial view　H. Tepals of staminate flower, abaxial view　I. Pistillate
flower, lateral view　J. Pistillate flower, adaxial view　K. Pistillate flower, abaxial view
L. Tepals of pistillate flower, adaxial view　M. Tepals of pistillate flower, abaxial view
N. Capsule　O. Ovary, transversely sectioned　P. Gynoecium, adaxial view　Q. Styles and
stigmas　R. Androecium, adaxial view

橙花侧膜秋海棠

Begonia aurantiflora C. I Peng, Yan Liu & S. M. Ku

多年生草本，具根状茎。叶互生，宽卵形或近圆形，基部心形，边缘具齿和短缘毛，先端钝至急尖；发育完全的成熟叶常在表面有带白色或银白色的环状带。花橙色；蒴果具不相等3翅；子房1室，侧膜胎座；花柱3。

分布： 原产于中国（广西）。

生境： 半荫蔽的石灰岩洞穴入口及内部的岩崖上。

Perennial herb, rhizomatous. Leaves alternate, broadly ovate or suborbicular, base cordate, margin denticulate and ciliolate, apex obtuse to acute; fully developed mature leaves often with a whitish or silver white ring-shaped belt on the adaxial surface. Flowers orange; capsule with unequal 3 wings; ovary 1-locular, placentation parietal; styles 3.

Distribution: Native to China (Guangxi).

Habitat: On semi-shady rocky cliffs at the entrance and inside of limestone caverns.

A

5cm

A. 植株　B. 花序　C. 叶片正反面　D. 叶片横切面　E. 雄花侧面观　F. 雄花反面观 G. 雄花花被片正面观　H. 雄花花被片反面观　I. 雌花侧面观　J. 雌花正面观　K. 雌 花花被片正面观　L. 雌花花被片反面观　M. 蒴果　N. 子房横切面　O. 雄蕊　P. 雌 蕊　Q. 花柱和柱头

A. Habit　B. Inflorescence　C. Leaf, adaxial and abaxial view　D. Leaf, transversely sectioned　E. Staminate flower, adaxial view　F. Staminate flower, abaxial view　G. Tepals of staminate flower, adaxial view　H. Tepals of staminate flower, abaxial view　I. Pistillate flower, lateral view　J. Pistillate flower, adaxial view　K. Tepals of pistillate flower, adaxial view　L. Tepals of pistillate flower, abaxial view　M. Capsule　N. Ovary, transversely sectioned　O. Androecium, lateral view　P. Gynoecium, adaxial view　Q. Styles and stigmas

桂南秋海棠

Begonia austroguangxiensis Y. M. Shui & W. H. Chen

多年生草本，具根状茎。叶基生，叶片宽卵形或近圆形，边缘有细锯齿和缘毛，先端圆形或钝形。花序腋生，二歧聚伞花序；花白色至粉红色，无毛；蒴果密被红色柔毛，具近等长3翅；子房1室，侧膜胎座；花柱3。

分布：原产于中国（广西）。

生境：开阔的石灰岩山林地。

Perennial herb, rhizomatous. Leaves basal, blades broadly ovate or suborbicular, margin serrulate and ciliate, apex rounded or obtuse. Inflorescences axillary, dichasial cyme; flowers pinkish or white, glabrous; capsule densely red villose, with subequal 3 wings; ovary 1-locular, placentation parietal; styles 3.

Distribution: Native to China (Guangxi).

Habitat: In the open forest on limestone hills.

A

5cm

A. 植株　B. 花序　C. 叶片正反面　D. 叶片横切面　E. 雄花正面观　F. 雄花反面观
G. 雄花花被片正面观　H. 雄花花被片反面观　I. 雌花侧面观　J. 雌花花被片正面观
K. 雌花花被片反面观　L. 蒴果　M. 雄蕊　N. 子房横切面　O. 雌蕊　P. 花柱和柱头
Q. 苞片

A. Habit　B. Inflorescence　C. Leaf, adaxial and abaxial view　D. Leaf, transversely
sectioned　E. Staminate flower, adaxial view　F. Staminate flower, abaxial view　G. Tepals
of staminate flower, adaxial view　H. Tepals of staminate flower, abaxial view　I. Pistillate
flower, lateral view　J. Tepals of pistillate flower, adaxial view　K. Tepals of pistillate flower,
abaxial view　L. Capsule　M. Androecium, lateral view　N. Ovary, transversely sectioned
O. Gynoecium, adaxial view　P. Styles and stigmas　Q. Bracts, adaxial and abaxial view

桂南秋海棠
Begonia austroguangxiensis Y. M. Shui & W. H. Chen

A

5cm

A. 植株　B. 花序　C. 叶片正反面　D. 雄花正面观　E. 雄花反面观　F. 雄花花被片正面观　G. 雄花花被片反面观　H. 雌花正面观　I. 雌花侧面观　J. 雌花花被片正面观　K. 雌花花被片反面观　L. 雄蕊　M. 蒴果　N. 子房横切面　O. 雌蕊　P. 花柱和柱头

A. Habit　B. Inflorescence　C. Leaf, adaxial and abaxial view　D. Staminate flower, adaxial view　E. Staminate flower, abaxial view　F. Tepals of staminate flower, adaxial view G. Tepals of staminate flower, abaxial view　H. Pistillate flower, adaxial view　I. Pistillate flower, lateral view　J. Tepals of pistillate flower, adaxial view　K. Tepals of pistillate flower, abaxial view　L. Androecium, lateral view　M. Capsule　N. Ovary, transversely sectioned O. Gynoecium, adaxial view　P. Styles and stigmas

西南秋海棠

Begonia cavaleriei H. Lév.

多年生草本，具根状茎。叶基生，叶柄长度可达30 cm；叶片盾状，卵形，先端渐尖，无毛。花白色至粉红色；蒴果长椭圆形，具不等长3翅；子房3室，中轴胎座；花柱3。

分布： 原产于中国（广西、贵州、云南）、越南。

生境： 林下的石灰岩上。

Perennial herb, rhizomatous. Leaves basal, petioles up to 30 cm; blades peltate, ovate, apex acuminate, glabrous. Flowers white to pinkish; capsule oblong, with unequal 3 wings; ovary 3-locular, placentae axile; styles 3.

Distribution: Native to China (Guangxi, Guizhou, Yunnan), and Vietnam.

Habitat: On limestone rocks in forests.

A
5cm

A. 植株　B. 花序　C. 叶片正反面和叶柄　D. 雄花正面观　E. 雄花反面观　F. 雄花花被片正面观　G. 雄花花被片反面观　H. 雌花正面观　I. 雌花侧面观　J. 雌花花被片正面观　K. 雌花花被片反面观　L. 雄蕊　M. 蒴果　N. 子房横切面　O. 雌蕊　P. 花柱和柱头　Q. 苞片

A. Habit　B. Inflorescence　C. Leaf, adaxial and abaxial view, and petiole　D. Staminate flower, adaxial view　E. Staminate flower, abaxial view　F. Tepals of staminate flower, adaxial view　G. Tepals of staminate flower, abaxial view　H. Pistillate flower, adaxial view　I. Pistillate flower, lateral view　J. Tepals of pistillate flower, adaxial view　K. Tepals of pistillate flower, abaxial view　L. Androecium, adaxial view　M. Capsule　N. Ovary, transversely sectioned　O. Gynoecium, adaxial view　P. Styles and stigmas　Q. Bracts, adaxial and abaxial view

西南秋海棠

Begonia cavaleriei H. Lév.

A

5cm

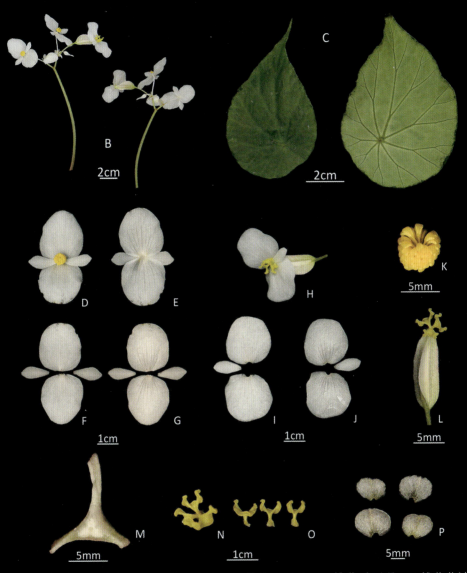

A. 植株　B. 花序　C. 叶片正反面　D. 雄花正面观　E. 雄花反面观　F. 雄花花被片正面观　G. 雄花花被片反面观　H. 雌花侧面观　I. 雌花花被片正面观　J. 雌花花被片反面观　K. 雄蕊　L. 蒴果　M. 子房横切面　N. 雌蕊　O. 花柱和柱头　P. 苞片

A. Habit　B. Inflorescence　C. Leaf, adaxial and abaxial view　D. Staminate flower, adaxial view　E. Staminate flower, abaxial view　F. Tepals of staminate flower, adaxial view G. Tepals of staminate flower, abaxial view　H. Pistillate flower, lateral view　I. Tepals of pistillate flower, adaxial view　J. Tepals of pistillate flower, abaxial view　K. Androecium, adaxial view　L. Capsule　M. Ovary, transversely sectioned　N. Gynoecium, adaxial view O. Styles and stigmas　P. Bracts, adaxial and abaxial view

双花秋海棠

Begonia biflora T. C. Ku

多年生草本，具根状茎，茎直立。叶片宽卵形；表面疏被短硬毛，背面被短柔毛。花序无毛或具柔毛；花淡黄绿色；蒴果具不等或近等长3翅；子房1室，侧膜胎座；花柱3。

分布：原产于中国（云南）。

生境：石灰岩山洞或林下。

Perennial herb, rhizomatous, stems erect. Leaf blades broadly ovate; adaxial surface sparsely hispidulous, abaxial surface pubescent. Inflorescences glabrous or pilose; flowers pale yellowish-green; capsule with unequal or subequal 3 wings; ovary 1-locular, placentation parietal; styles 3.

Distribution: Native to China (Yunnan).

Habitat: In limestone caves or forests.

A

5cm

A. 植株　B. 花序　C. 叶片正反面　D. 叶片横切面　E. 雄花正面观　F. 雄花反面观 G. 雄花花被片正面观　H. 雄花花被片反面观　I. 雌花正面观　J. 雌花反面观　K. 雌花花被片正面观　L. 雌花花被片反面观　M. 雄蕊　N. 蒴果　O. 子房横切面　P. 雌蕊 Q. 花柱和柱头　R. 苞片

A. Habit　B. Inflorescence　C. Leaf, adaxial and abaxial view　D. Leaf, transversely sectioned　E. Staminate flower, adaxial view　F. Staminate flower, abaxial view　G. Tepals of staminate flower, adaxial view　H. Tepals of staminate flower, abaxial view　I. Pistillate flower, adaxial view　J. Pistillate flower, abaxial view　K. Tepals of pistillate flower, adaxial view　L. Tepals of pistillate flower, abaxial view　M. Androecium, adaxial view　N. Capsule O. Ovary, transversely sectioned　P. Gynoecium, adaxial view　Q. Styles and stigmas R. Bracts, adaxial and abaxial view

崇左秋海棠

Begonia chongzuoensis Yan Liu, S. M. Ku & C. I Peng

多年生草本，根状茎红褐色。叶片宽卵形，基部心形，前端尖锐，纸质，边缘呈锯齿状且具缘毛；表面通常深棕色至紫红色，被稀疏的硬毛，稍具皱；叶柄棕红色，被毛。花白色或稍显粉色；蒴果三棱椭圆形，无毛，具等长或近等长3翅；子房1室，侧膜胎座；花柱3。

分布： 原产于中国（广西）。

生境： 常绿森林中陡峭的石灰岩斜坡下。

Perennial herb, rhizomes reddish brown. Leaves broadly ovate, base deeply cordate, apex acuminate or, texture papery, margin crenate-denticulate and ciliate; adaxial surface shortly dark brownish to purplish red, sparsely covered with coarse hairs, slightly rugose; petioles brownish red, sparsely hirsute-villous. Flowers white or somewhat pinkish; capsule trigonous-ellipsoid, glabrous, with equal or subequal 3 wings; 1-locular, placentation parietal; styles 3.

Distribution: Native to China (Guangxi).

Habitat: Base of steep limestone slopes in an evergreen forest.

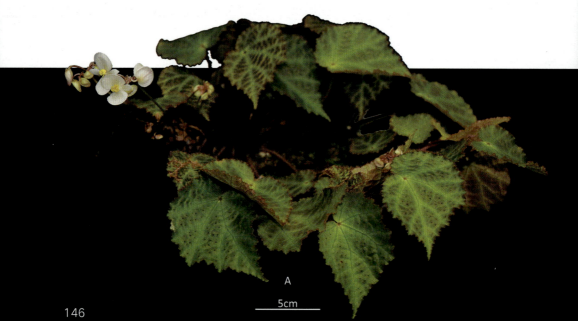

A

5cm

A. 植株　B. 花序　C. 叶片正反面和叶柄　D. 叶片横切面　E. 雌花正面观　F. 雌花侧面观　G. 雌花花被片正面观　H. 雌花花被片反面观　I. 雄花正面观　J. 雄花反面观　K. 雄花花被片正面观　L. 雄花花被片反面观　M. 蒴果　N. 子房横切面　O. 雌蕊　P. 花柱和柱头　Q. 雄蕊

A. Habit　B. Inflorescence　C. Leaf, adaxial and abaxial view, and petiole　D. Leaf, transversely sectioned　E. Pistillate flower, adaxial view　F. Pistillate flower, lateral view　G. Tepals of pistillate flower, adaxial view　H. Tepals of pistillate flower, abaxial view　I. Staminate flower, adaxial view　J. Staminate flower, abaxial view　K. Tepals of staminate flower, adaxial view　L. Tepals of staminate flower, abaxial view　M. Capsule　N. Ovary, transversely sectioned　O. Gynoecium, adaxial view　P. Styles and stigmas　Q. Androecium, lateral view

卷毛秋海棠

Begonia cirrosa L. B. Sm. & Wassh.

多年生草本，根状茎伸长。叶基生，叶片宽卵形至近圆形，纸质；表面具短硬毛，背面浅绿色或红色。花白色或粉红色；蒴果椭圆形，被红色硬毛，具不等长3翅；子房1室，侧膜胎座；花柱3。

分布： 原产于中国（云南、广西）。

生境： 海拔1 000 m左右的石灰岩林下的荫蔽岩石上。

Perennial herb, rhizomes elongate. Leaves basal, blades broadly ovate to suborbicular, papery; adaxial surface setose-pilose, tertiary veins weakly percurrent, abaxial surface pale green or reddish, pilose or hirsute-pilose. Flowers white or pinkish; capsule oblong, red hispid-hirsute, with unequal 3 wings; ovary 1-locular, placentation parietal; styles 3.

Distribution: Native to China (Yunnan, Guangxi).

Habitat: Limestone forests, on rocks in shaded environments, *ca.* 1,000 m elevation.

A

5cm

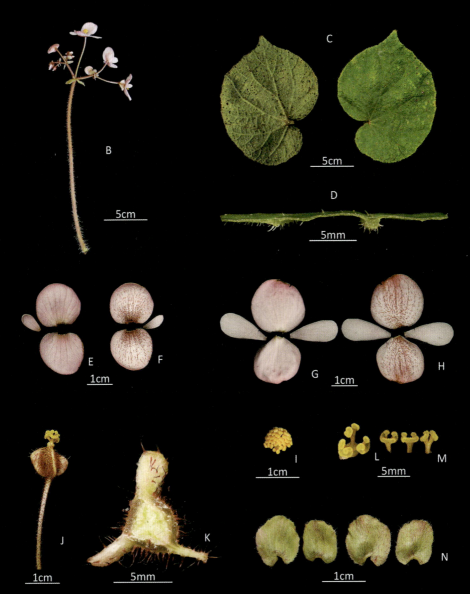

A. 植株　B. 花序　C. 叶片正反面　D. 叶片横切面　E. 雄花花被片正面观　F. 雄花花被片反面观　G. 雌花花被片正面观　H. 雌花花被片反面观　I. 雄蕊　J. 蒴果　K. 子房横切面　L. 雌蕊　M. 花柱和柱头　N. 苞片

A. Habit　B. Inflorescence　C. Leaf, adaxial and abaxial view　D. Leaf, transversely section　E. Tepals of staminate flower, adaxial view　F. Tepals of staminate flower, abaxial view　G. Tepals of pistillate flower, adaxial view　H. Tepals of pistillate flower, abaxial view　I. Androecium, lateral view　J. Capsule　K. Ovary, transversely sectioned　L. Gynoecium, adaxial view　M. Styles and stigmas　N. Bracts, adaxial and abaxial view

柱果秋海棠

Begonia cylindrica D. R. Liang & X. X. Chen

多年生草本，根状茎密被褐色卵状披针形膜质鳞片。叶盾形，叶片宽卵形或近圆形，先端有短尾尖。花白色或粉红色；雄花花被片4，雌花花被片3；蒴果长圆筒状，无毛，无翅；子房3室，中轴胎座；花柱3。

分布： 原产于中国（广西）。

生境： 林下阴湿环境的石灰岩上。

Perennial herb, rhizomes with densely brown ovate lanceolate membranous scales. Leaves peltate, blades broadly ovate or suborbicular, apex short caudal tip. Flowers white or pink; male flowers petals 4, female flowers petals 3; capsule long cylindric, glabrous, wingless; ovary 3-locular, placentae axial; styles 3.

Distribution: Native to China (Guangxi).

Habitat: Forests, on limestone rocks in shaded moist environments.

A

5cm

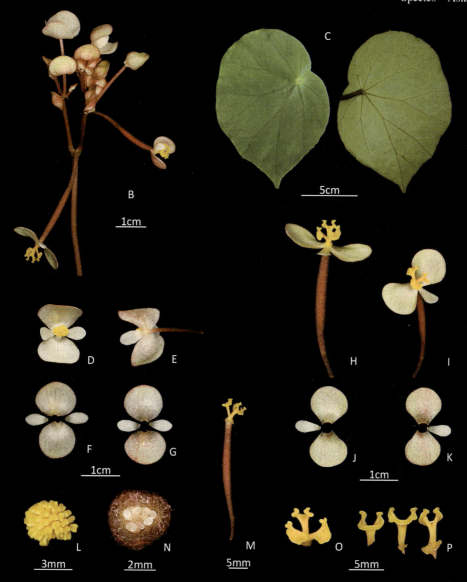

A. 植株　B. 花序　C. 叶片正反面　D. 雄花正面观　E. 雄花反面观　F. 雄花花被片正面观　G. 雄花花被片反面观　H. 雌花侧面观　I. 雌花正面观　J. 雌花花被片正面观　K. 雌花花被片反面观　L. 雄蕊　M. 蒴果　N. 子房横切面　O. 雌蕊　P. 花柱和柱头

A. Habit　B. Inflorescence　C. Leaf, adaxial and abaxial view　D. Staminate flower, adaxial view　E. Staminate flower, abaxial view　F. Tepals of staminate flower, adaxial view G. Tepals of staminate flower, abaxial view　H. Pistillate flower, lateral view　I. Pistillate flower, adaxial view　J. Tepals of pistillate flower, adaxial view　K. Tepals of pistillate flower, abaxial view　L. Androecium, adaxial view　M. Capsule　N. Ovary, transversely sectioned　O. Gynoecium, adaxial view　P. Styles and stigmas

方氏秋海棠

Begonia fangii Y. M. Shui & C. I Peng

多年生草本，具匍匐茎，高约40 cm。掌状复叶基生，小叶披针形；表面深绿色，背面红褐色。花序花序稀疏至中等被毛；花粉色；蒴果近无毛，具近等长3翅；子房1室，侧膜胎座；花柱3。

分布：原产于中国（广西）。

生境：林下石灰岩岩壁上。

Perennial herb, prostrate, up to *ca*. 40 cm tall. Leaves palmately compound basal, blades lanceolate; adaxial surface dark green, abaxial surface reddish brown. Inflorescences sparsely to moderately pubescent; flowers pinkish; capsule subglabrous, with subequal 3 wings; ovary 1-locular, placentation parietal; styles 3.

Distribution: Native to China (Guangxi).

Habitat: On limestone rocks in forests.

A

10cm

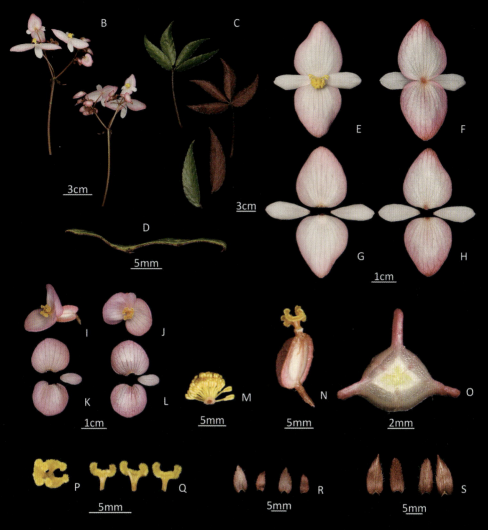

A. 植株　B. 花序　C. 叶片正反面　D. 叶片横切面　E. 雄花正面观　F. 雄花反面观　G. 雄花花被片正面观　H. 雄花花被片反面观　I. 雌花侧面观　J. 雌花正面观　K. 雌花花被片正面观　L. 雌花花被片反面观　M. 雄蕊　N. 蒴果　O. 子房横切面　P. 雌蕊　Q. 花柱和柱头　R. 小苞片　S. 苞片

A. Habit　B. Inflorescence　C. Leaf, adaxial and abaxial view　D. Leaf, transversely sectioned　E. Staminate flower, adaxial view　F. Staminate flower, abaxial view　G. Tepals of staminate flower, adaxial view　H. Tepals of staminate flower, abaxial view　I. Pistillate flower, lateral view　J. Pistillate flower, adaxial view　K. Tepals of pistillate flower, adaxial view　L. Tepals of pistillate flower, abaxial view　M. Androecium, lateral view　N. Capsule O. Ovary, transversely sectioned　P. Gynoecium, adaxial view　Q. Styles and stigmas R. Bracteoles, adaxial and abaxial view　S. Bracts, adaxial and abaxial view

黑峰秋海棠

Begonia ferox C. I Peng & Yan Liu

多年生草本，具根状茎。叶基生和茎生，叶片卵形，先端渐尖；表面泡状突起，单个突起呈锥形，顶端红色。花黄红色或黄绿色；蒴果三棱椭圆形，具不等长3翅，背翅新月形，侧翅狭窄；子房1室，侧膜胎座；花柱3。

分布： 原产于中国（广西）。

生境： 常绿阔叶林中落叶丰富的石灰岩或裸露的岩坡上。

Perennial herb, rhizomatous. Leaves basal and cauline, blades ovate, apex acuminate; adaxial surface bullate, individual bullae conical, tip reddish. Flowers reddish-yellow or greenish-yellow; capsule trigonous-ellipsoid, with unequal 3 wings, abaxial wing crescent-shaped, lateral wings narrowed; ovary 1-locular, placentation parietal; styles 3.

Distribution: Native to China (Guangxi).

Habitat: On limestone rocks with abundant leaf litter or on bare rocky slopes in evergreen broadleaf forests.

A
5cm

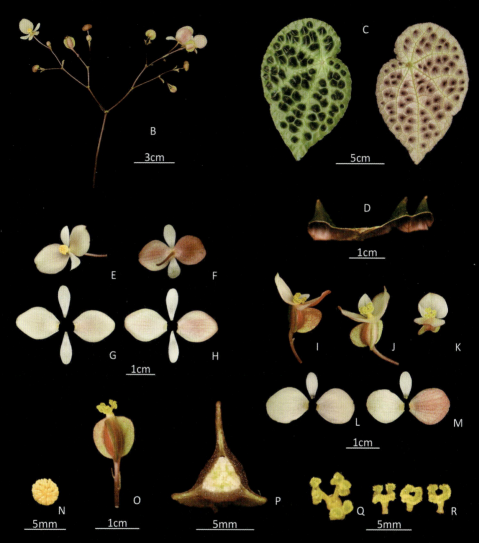

A. 植株　B. 花序　C. 叶片正反面　D. 叶片横切面　E. 雄花正面观　F. 雄花反面观
G. 雄花花被片正面观　H. 雄花花被片反面观　I, J. 雌花侧面观　K. 雌花正面观
L. 雌花花被片正面观　M. 雌花花被片反面观　N. 雄蕊　O. 蒴果　P. 子房横切面
Q. 雌蕊　R. 花柱和柱头

A. Habit　B. Inflorescence　C. Leaf, adaxial and abaxial view　D. Leaf, transversely
sectioned　E. Staminate flower, adaxial view　F. Staminate flower, abaxial view　G. Tepals
of staminate flower, adaxial view　H. Tepals of staminate flower, abaxial view　I, J. Pistillate
flower, lateral view　K. Pistillate flower, adaxial view　L. Tepals of pistillate flower, adaxial view
M. Tepals of pistillate flower, abaxial view　N. Androecium, adaxial view　O. Capsule　P. Ovary,
transversely sectioned　Q. Gynoecium, adaxial view　R. Styles and stigmas

丝形秋海棠

Begonia filiformis Irmsch.

多年生草本，根状茎匍匐。叶片斜卵形或近圆形，表面被绵状的柔毛，主脉间呈浓绿色，有白色斑点。总状花序，花梗密被具硬腺毛；花淡绿色；雄花花被片4，雌花3；蒴果椭圆形，被腺毛，具不等长3翅；子房1室，侧膜胎座；花柱3。

分布：原产于中国（广西）、越南北部。

生境：岩石表面或石灰岩崖壁上。

Perennial herb, rhizomes creeping. Leaf blades obliquely ovate to suborbicular, abaxial surface woolly-villous, intense green with white spots between the main veins. Inflorescences racemose, peduncles densely glandular hispid; flowers pale green; male flowers tepals 4, female flowers tepals 3; capsule elliptic, glandular hispid, with unequal 3 wings; ovary 1-locular, placentation parietal; styles 3.

Distribution: Native to China (Guangxi), and Northern Vietnam.

Habitat: On rocky surfaces or limestone cliffs.

A

5cm

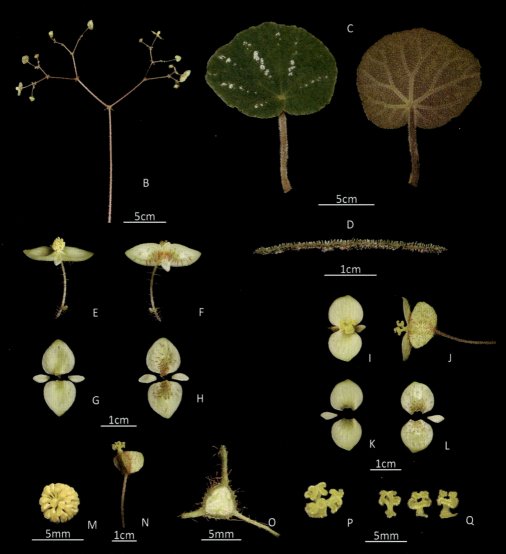

A. 植株　B. 花序　C. 叶片正反面　D. 叶片横切面　E. 雄花侧面观　F. 雄花反面观 G. 雄花花被片正面观　H. 雄花花被片反面观　I. 雌花正面观　J. 雌花侧面观　K. 雌花花被片正面观　L. 雌花花被片反面观　M. 雄蕊　N. 蒴果　O. 子房横切面　P. 雌蕊 Q. 花柱和柱头

A. Habit　B. Inflorescence　C. Leaf, adaxial and abaxial view　D. Leaf, transversely sectioned　E. Staminate flower, lateral view　F. Staminate flower, abaxial view　G. Tepals of staminate flower, adaxial view　H. Tepals of staminate flower, abaxial view　I. Pistillate flower, adaxial view　J. Pistillate flower, lateralview　K. Tepals of pistillate flower, adaxial view L. Tepals of pistillate fl ower, abaxial view　M. Androecium, adaxial view　N. Capsules　O. Ovary, transversely sectioned　P. Gynoecium, adaxial view　Q. Styles and stigmas

须苞秋海棠

Begonia fimbribracteata Y. M. Shui & W. H. Chen

多年生草本，高10～30 cm，无直立茎。叶基生，叶片宽卵形，先端圆形；具皱纹和圆锥状乳突，边缘有缘毛和细锯齿。花白色；雄花花被片4，雌花花被片3；蒴果三棱椭圆形，具近等长3翅；子房1室，侧膜胎座；花柱3。

分布： 原产于中国（广西）。

生境： 石灰岩林下岩壁。

Perennial herb, 10-30 cm tall, without erect stems. Leaves basal, blades broadly ovate, apex rotundate, with wrinkled and conical papillae, margin ciliate and serrulate. Flowers white; male flowers tepals 4, female flowers tepals 3; capsule trigonous-ellipsoid, with subequal 3 wings; ovary 1-locular, placentation parietal; styles 3.

Distribution: Native to China (Guangxi).

Habitat: On limestone rocks in forests.

A

5cm

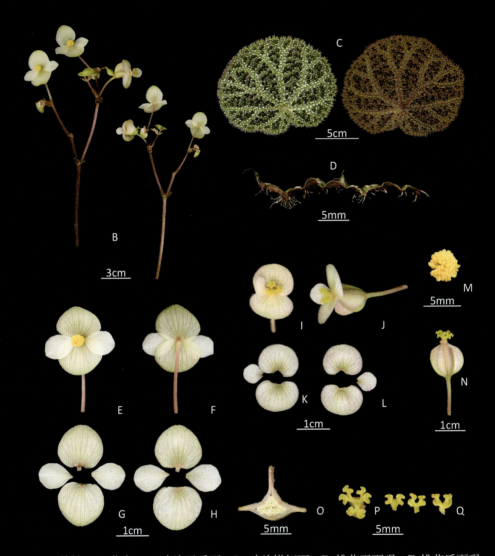

A. 植株　B. 花序　C. 叶片正反面　D. 叶片横切面　E. 雄花正面观　F. 雄花反面观
G. 雄花花被片正面观　H. 雄花花被片反面观　I. 雌花正面观　J. 雌花侧面观　K. 雌花花被片正面观　L. 雌花花被片反面观　M. 雄蕊　N. 蒴果　O. 子房横切面　P. 雌蕊
Q. 花柱和柱头

A. Habit　B. Inflorescence　C. Leaf, adaxial and abaxial view　D. Leaf, transversely sectioned　E. Staminate flower, adaxial view　F. Staminate flower, abaxial view　G. Tepals of staminate flower, adaxial view　H. Tepals of staminate flower, abaxial view　I. Pistillate flower, adaxial view　J. Pistillate flower, lateral view　K. Tepals of pistillate flower, adaxial view　L. Tepals of pistillate flower, abaxial view　M. Androecium, adaxial view　N. Capsule O. Ovary, transversely sectioned　P. Gynoecium, adaxial view　Q. Styles and stigmas

广东秋海棠

Begonia guangdongensis W. H. Tu, B. M. Wang & L. Y. Li

多年生草本，具根状茎。叶基生，叶片斜卵形，具皱；表面密被白色长硬毛，叶脉凹陷，背面叶脉凸起；叶柄红色，密被白色柔毛。花粉色；雄花花被片4，雌花花被片3；蒴果三棱椭圆形，深粉红色，无毛，具不等长3翅；子房1室，侧膜胎座；花柱3。

分布： 原产于中国（广东）。

生境： 海拔80～100 m的常绿林中的石灰岩山的斜坡上。

Perennial herb, rhizomatous. Leaves basal, blades obliquely ovate, rugose; adaxial surface densely white hirsute-pilose, veins depressed, abaxial surface veins convex; petioles red, with densely white villose; Flowers pinkish; male flowers tepals 4, female flowers tepals 3; capsule trigonous-ellipsoid, dark pink, glabrous, with unequal 3 wings; ovary 1-locular, placentation parietal; styles 3.

Distribution: Native to China (Guangdong).

Habitat: On the slope of a limestone hill in evergreen forests, 80-100 m elevation.

A

10cm

A. 植株　B. 花序　C. 叶片正反面　D. 叶片横切面　E. 雄花正面观　F. 雄花反面观　G. 雄花花被片正面观　H. 雄花花被片反面观　I. 雌花正面观　J. 雌花侧面观　K. 雌花花被片正面观　L. 雌花花被片反面观　M. 雄蕊　N. 蒴果　O. 子房横切面　P. 雌蕊　Q. 花柱和柱头

A. Habit　B. Inflorescence　C. Leaf, adaxial and abaxial view　D. Leaf, transversely sectioned　E. Staminate flower, adaxial view　F. Staminate flower, abaxial view　G. Tepals of staminate flower, adaxial view　H. Tepals of staminate flower, abaxial view　I. Pistillate flower, adaxial view　J. Pistillate flower, lateral view　K. Tepals of pistillate flower, adaxial view　L. Tepals of pistillate flower, abaxial view　M. Androecium, adaxial view　N. Capsule　O. Ovary, transversely sectioned　P. Gynoecium, adaxial view　Q. Styles and stigmas

丽纹秋海棠

Begonia kui C. I Peng

多年生草本，根茎类，全株多毛。该种在外观上与彩纹秋海棠（*Begonia variegata*）相似，但其叶片顶端相对较圆润，边缘有白色斑点。花序有毛；外轮花被片呈玫瑰色或红色，背面被短毛；蒴果无毛。

分布：具体原产地不详，可能产自越南北部。

生境：落叶丰富的石灰岩表面。

Perennial herb, rhizomatous, hairy throughout. The species is similar in appearance to *Begonia variegata*, but differs by having leaves with obtuse or rounded apices, margins with white spots. Inflorescences hairy; outer tepals rose or reddish with short hairs on abaxial surface; capsule glabrous.

Distribution: Original locality not known, possibly from Northern Vietnam.

Habitat: On limestone rock surface with abundant leaf litter.

A

5cm

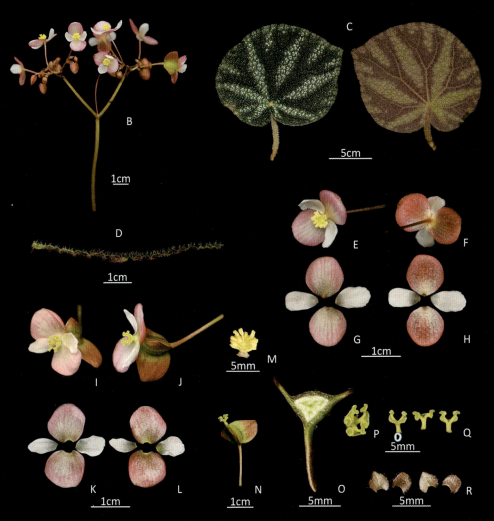

A. 植株　B. 花序　C. 叶片正反面　D. 叶片横切面　E. 雄花正面观　F. 雄花反面观
G. 雄花花被片正面观　H. 雄花花被片反面观　I. 雌花正面观　J. 雌花侧面观　K. 雌
花花被片正面观　L. 雌花花被片反面观　M. 雄蕊　N. 蒴果　O. 子房横切面　P. 雌蕊
Q. 花柱和柱头　R. 苞片

A. Habit　B. Inflorescence　C. Leaf, adaxial and abaxial view　D. Leaf, transversely
sectioned　E. Staminate flower, adaxial view　F. Staminate flower, abaxial view　G. Tepals
of staminate flower, adaxial view　H. Tepals of staminate flower, abaxial view　I. Pistillate
flower, adaxial view　J. Pistillate flower, lateral view　K. Tepals of pistillate flower, adaxial
view　L. Tepals of pistillate flower, abaxial view　M. Androecium, lateral view　N. Capsule
O. Ovary, transversely sectioned　P. Gynoecium, adaxial view　Q. Styles and stigmas
R. Bracts, adaxial and abaxial view

灯果秋海棠

Begonia lanternaria Irmsch.

多年生草本，具根状茎。叶基生，叶柄疏生长柔毛；叶片卵形至宽卵形；表面淡褐绿色，无毛，革质，先端短尾状，背面叶脉蜘蛛网状。花白色至粉色；蒴果无毛，具不等长3翅；子房1室，侧膜胎座；花柱3。

分布： 原产于中国（广西、云南）、越南北部。

生境： 森林边缘的石灰岩上。

Perennial herb, rhizomatous. Leaves basal, petioles sparsely villous; blades ovate to broadly ovate; adaxial surface pale brown-green, glabrous, leathery, apex shortly caudate, abaxial surface venation web-like. Flowers white to pinkish; capsule glabrous, with unequal 3 wings; ovary 1-locular, placentation parietal; styles 3.

Distribution: Native to China (Guangxi, Yunnan), and Northern Vietnam.

Habitat: On limestone rocks at forest margins.

A

5cm

A. 植株　B. 花序　C. 叶片正反面　D. 叶片横切面　E. 雄花正面观　F. 雄花反面观 G. 雄花花被片正面观　H. 雄花花被片反面观　I. 雌花正面观　J. 雌花侧面观　K. 雌花 花被片正面观　L. 雌花花被片反面观　M. 雄蕊　N. 蒴果　O. 子房横切面　P. 雌蕊 Q. 花柱和柱头

A. Habit　B. Inflorescence　C. Leaf, adaxial and abaxial view　D. Leaf, transversely sectioned　E. Staminate flower, adaxial view　F. Staminate flower, abaxial view　G. Tepals of staminate flower, adaxial view　H. Tepals of staminate flower, abaxial view　I. Pistillate flower, adaxial view　J. Pistillate flower, lateral view　K. Tepals of pistillate flower, adaxial view　L. Tepals of pistillate flower, abaxial view　M. Androecium, adaxial view　N. Capsule O. Ovary, transversely sectioned　P. Gynoecium, adaxial view　Q. Styles and stigmas

癞叶秋海棠

Begonia leprosa Hance

多年生草本，具根状茎。叶柄幼时无毛到密被棕色长柔毛；叶片有时呈非常浅的盾状，近圆形，圆形或宽卵形，边缘略具细锯齿，先端急尖或短尾状。花白色至粉红色，无毛；雄花花被片4，雌花花被片3；蒴果长圆筒状，无毛，无翅；子房3室，中轴胎座；柱头3。

分布： 原产于中国（广东、广西）。

生境： 半开阔森林或灌丛中石灰岩上。

Perennial herb, rhizomatous. Petioles glabrous to densely brown villous when young; leaf blades sometimes very shallowly peltate, suborbicular, obovate, or broadly ovate, margin remotely and minutely serrulate, apex acute or shortly caudate. Flowers white to pink, glabrous; male tepals 4, female tepals 3; capsule long cylindric, glabrous, wingless; ovary 3-locular, placentae axile; styles 3.

Distribution: Native to China (Guangdong, Guangxi).

Habitat: Semi-open forests or scrubby vegetation, on limestone rocks.

A

10cm

A. 植株　B. 花序　C. 叶片正反面和叶柄　D. 叶片横切面　E. 雄花正面观　F. 雄花反面观　G. 雄花花被片正面观　H. 雄花花被片反面观　I. 雌花侧面观　J. 雌花花被片正面观　K. 雌花花被片反面观　L. 雄蕊　M. 蒴果　N. 子房横切面　O. 雌蕊　P. 花柱和柱头　Q. 苞片

A. Habit　B. Inflorescence　C. Leaf, adaxial and abaxial view, and petiole　D. Leaf, transversely sectioned　E. Staminate flower, adaxial view　F. Staminate flower, abaxial view
G. Tepals of staminate flower, adaxial view　H. Tepals of staminate flower, abaxial view
I. Pistillate flower, lateral view　J. Tepals of pistillate flower, adaxial view　K. Tepals of pistillate flower, abaxial view　L. Androecium, lateral view　M. Capsule　N. Ovary, transversely sectioned
O. Gynoecium, lateral view　P. Styles and stigmas　Q. Bracts, adaxial and abaxial view

刘演秋海棠

Begonia liuyanii C. I Peng, S. M. Ku & W. C. Leong

多年生草本，具根状茎。叶片近革质，宽卵形或近圆形，基部斜心形，先端急尖；表面绿色或深绿色，略具皱，疏生刚毛，背面色浅（幼嫩时略带红色），被绵毛。花黄绿色或带红色；雄花花被片4，雌花花被片3；蒴果三棱椭圆形，红色，被红色硬腺毛，具不等长3翅；子房1室，侧膜胎座；柱头3。

分布： 原产于中国（广西）。

生境： 荫蔽的丘陵山地的森林石灰岩斜坡上。

Perennial herb, rhizomatous. Leaves subcoriaceous, blades broadly ovate or suborbicular, base strongly oblique-cordate, apex shortly acuminate; adaxial surface green or dark green, slightly rugose, sparsely setose, abaxial surface pale (reddish when young), lanuginous. Flowers yellowish or reddish; male flowers tepals 4, female flowers tepals 3; capsule trigonous-ellipsoid, reddish, red glandular-hispid, with unequal 3 wings; ovary 1-locular, placentation parietal; styles 3.

Distribution: Native to China (Guangxi).

Habitat: Broadleaved forests; on shaded, rocky limestone slopes.

A

10cm

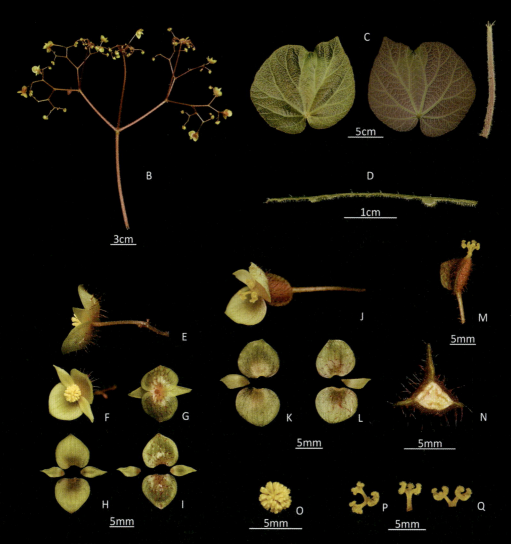

A. 植株　B. 花序　C. 叶片正反面　D. 叶片横切面　E. 雄花侧面观　F. 雄花正面观 G. 雄花反面观　H. 雄花花被片正面观　I. 雄花花被片反面观　J. 雌花侧面观　K. 雌花花被片正面观　L. 雌花花被片反面观　M. 蒴果　N. 子房横切面　O. 雄蕊　P. 雌蕊 Q. 花柱和柱头

A. Habit　B. Inflorescence　C. Leaf, adaxial and abaxial view　D. Leaf, transversely sectioned　E. Staminate flower, lateral view　F. Staminate flower, adaxial view　G. Staminate flower, abaxial view　H. Tepals of staminate flower, adaxial view　I. Tepals of staminate flower, abaxial view　J. Pistillate flower, lateral view　K. Tepals of pistillate flower, adaxial view　L. Tepals of pistillate flower, abaxial view　M. Capsule　N. Ovary, transversely sectioned　O. Androecium, adaxial view　P. Gynoecium, adaxial view　Q. Styles and stigmas

弄岗秋海棠

Begonia longgangensis C. I Peng & Yan Liu

多年生草本，最初生长出根茎，后变成匍匐茎，匍匐茎可达150 cm或更长。叶互生，叶柄圆柱状；叶片宽卵形到近圆形，先端渐尖，近革质；表面绿色，疏生红色微糙毛。花白色至粉红色；蒴果三棱椭圆形，具不等长3翅；子房1室，侧膜胎座；柱头3。

分布：原产于中国（广西）。

生境：常绿阔叶林锯齿状石灰岩上。

Perennial herb, rhizomatous initially and becoming stoloniferous, stolons to 150 cm or longer. Leaves alternate, petioles terete; blades broadly ovate to suborbicular, apex acuminate, subcoriaceous; adaxial surface green, sparsely reddish-scabridulous. Flowers white to pinkish; capsule trigonous-ellipsoid, with unequal 3 wings; ovary 1-locular, placentation parietal; styles 3.

Distribution: Native to China (Guangxi).

Habitat: On jagged limestone rocks in evergreen broadleaf forest.

A. 植株　B. 花序　C. 叶片正反面　D. 叶片横切面　E. 雄花正面观　F. 雄花反面观 G. 雄花花被片正面观　H. 雄花花被片反面观　I. 雌花正面观　J. 雌花侧面观　K. 雌花 花被片正面观　L. 雌花花被片反面观　M. 雄蕊　N. 蒴果　O. 子房横切面　P. 雌蕊 Q. 花柱和柱头

A. Habit　B. Inflorescence　C. Leaf, adaxial and abaxial view　D. Leaf, transversely sectioned　E. Staminate flower, adaxial view　F. Staminate flower, abaxial view　G. Tepals of staminate flower, adaxial view　H. Tepals of staminate flower, abaxial view　I. Pistillate flower, adaxial view　J. Pistillate flower, lateral view　K. Tepals of pistillate flower, adaxial view　L. Tepals of pistillate flower, abaxial view　M. Androecium, adaxial view　N. Capsule O. Ovary, transversely sectioned　P. Gynoecium, adaxial view　Q. Styles and stigmas

陆氏秋海棠

Begonia lui S. M. Ku, C. I Peng & Yan Liu

多年生草本，具根状茎。叶互生，叶片宽卵形或近圆形，基部斜心形；表面略具皱，深绿色，叶脉间有白色斑点，密被细刚毛，背面沿主脉和细脉被绒毛。花被片粉白色；蒴果具不等长3翅；子房1室，侧膜胎座；花柱3。

分布：原产于中国（广西）。

生境：石灰岩山上。

Perennial herb, rhizomatous. Leaves alternate, blades broadly ovate or suborbicular, base strongly oblique-cordate; adaxial surface slightly rugose, deep green, with white maculation in intercostal areas, moderately densely setulose, abaxial surface tomentose along veins and veinlets. Flowers pinkish white; capsule with unequal 3 wings; ovary 1-locular, placentation parietal; styles 3.

Distribution: Native to China (Guangxi).

Habitat: On limestone hills.

A

5cm

A. 植株　B. 花序　C. 叶片正反面　D. 雄花正面观　E. 雄花反面观　F. 雄花花被片正面观　G. 雄花花被片反面观　H. 雌花侧面观　I. 雌花反面观　J. 雌花花被片正面观　K. 雌花花被片反面观　L. 雄蕊　M. 蒴果　N. 子房横切面　O. 雌蕊　P. 花柱和柱头

A. Habit　B. Inflorescence　C. Leaf, adaxial and abaxial view　D. Staminate flower, adaxial view　E. Staminate flower, abaxial view　F. Tepals of staminate flower, adaxial view G. Tepals of staminate flower, abaxial view　H. Pistillate flower, lateral view　I. Pistillate flower, abaxial view　J. Tepals of pistillate flower, adaxial view　K. Tepals of pistillate flower, abaxial view　L. Androecium, adaxial view　M. Capsule　N. Ovary, transversely sectioned　O. Gynoecium, adaxial view　P. Styles and stigmas

陆氏秋海棠

Begonia lui S. M. Ku, C. I Peng & Yan Liu

A

5cm

A. 植株　B. 花序　C. 叶片正反面　D. 叶片横切面　E. 雄花正面观　F. 雄花反面观 G. 雄花花被片正面观　H. 雄花花被片反面观　I. 雌花正面观　J. 雌花侧面观　K. 雌花 花被片正面观　L. 雌花花被片反面观　M. 蒴果　N. 子房横切面　O. 雄蕊　P. 雌蕊 Q. 花柱和柱头

A. Habit　B. Inflorescence　C. Leaf, adaxial and abaxial view　D. Leaf, transversely sectioned　E. Staminate flower, adaxial view　F. Staminate flower, abaxial view　G. Tepals of staminate flower, adaxial view　H. Tepals of staminate flower, abaxial view　I. pistillate flower, adaxial view　J. Pistillate flower, lateral view　K. Tepals of pistillate flower, adaxial view　L. Tepals of pistillate flower, abaxial view　M. Capsule　N. Ovary, transversely sectioned　O. Androecium, lateral view　P. Gynoecium, adaxial view　Q. Styles and stigmas

罗城秋海棠

Begonia luochengensis S. M. Ku, C. I Peng & Yan Liu

多年生草本，具根状茎。叶互生，叶片斜卵形，表面具紫褐色斑纹，沿主脉形成白色条纹，靠近主要脉和侧脉附近呈淡绿色。花粉红色；蒴果无毛，具不等长3翅，花期呈粉红色；子房1室，侧膜胎座；柱头3。

分布： 原产于中国（广西）。

生境： 半遮阴、干燥或稍潮湿的石灰岩山上。

Perennial herb, rhizomatous. Leaves alternate, blades obliquely ovate, white maculations on adaxial surface, often along the midrib, forming a white zone, pale green near the major veins and major lateral veins. Flowers pinkish; ovary glabrous, with unequal 3 wings, pinkish at anthesis; ovary 1-locular, placentation parietal; styles 3.

Distribution: Native to China (Guangxi).

Habitat: On semi-shaded, dry or slightly moist limestone hills.

A

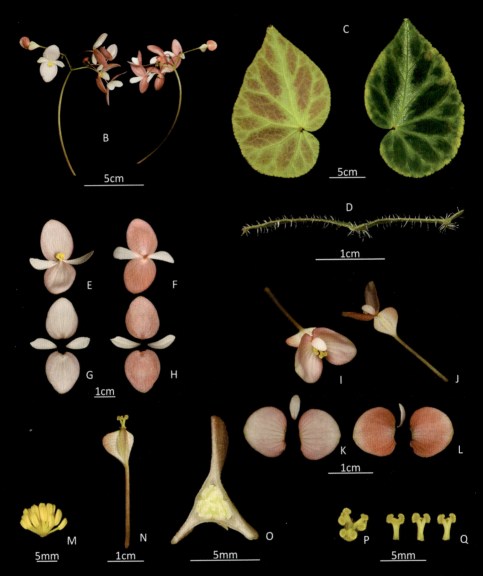

A. 植株　B. 花序　C. 叶片正反面　D. 叶片横切面　E. 雄花正面观　F. 雄花反面观
G. 雄花花被片正面观　H. 雄花花被片反面观　I, J. 雌花侧面观　K. 雌花花被片正面观
L. 雌花花被片反面观　M. 雄蕊　N. 蒴果　O. 子房横切面　P. 雌蕊　Q. 花柱和柱头

A. Habit　B. Inflorescence　C. Leaf, adaxial and abaxial view　D. Leaf, transversely
sectioned　E. Staminate flower, adaxial view　F. Staminate flower, abaxial view　G. Tepals
of staminate flower, adaxial view　H. Tepals of staminate flower, abaxial view　I, J. Pistillate
flower, lateral view　K. Tepals of pistillate flower, adaxial view　L. Tepals of pistillate flower,
abaxial view　M. Androecium, lateral view　N. Capsule　O. Ovary, transversely sectioned
P. Gynoecium, adaxial view　Q. Styles and stigmas

鹿寨秋海棠

Begonia luzhaiensis T. C. Ku

A1

10cm

A2

10cm

A3

10cm

多年生草本，根状茎宿存褐色鳞片。叶片宽卵形至近圆形，先端尾状渐尖，基部深心形；表面深绿色，大部分居群具紫褐色斑纹。花白色至粉色；雄花花被片4，雌花花被片3；蒴果无毛，具不等长3翅；子房1室，侧膜胎座；柱头3。

分布：原产于中国（广西）。

生境：有光线透入的洞穴、石灰岩山上岩石斜坡上。

Perennial herb, rhizomes with residual brown scales. Leaf blades broadly ovate to suborbicular, apex caudate acuminate, base deeply cordate; adaxial surface dark green, with purple-brown markings in most populations. Flowers white to pink; male flowers tepals 4, female flowers tepals 3; capsule glabrous, with unequal 3 wings; ovary 1-locular, placentation parietal; styles 3.

Distribution: Native to China (Guangxi).

Habitat: On limestone hills, rocky slopes, and in caves with some penetrating light.

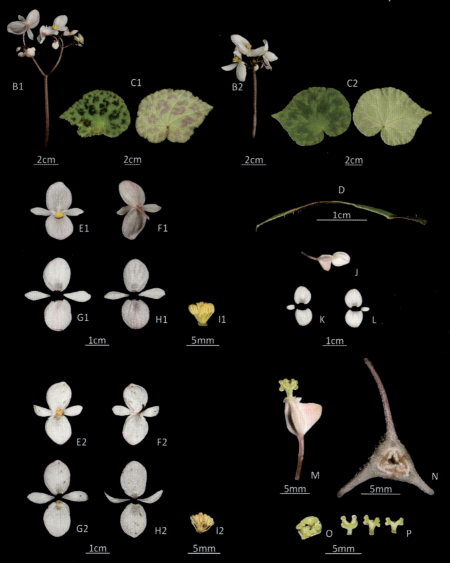

A1-A3. 植株　B1，B2. 花序　C1，C2. 叶片正反面　D. 叶片横切面　E1，E2. 雄花正面观　F1，F2. 雄花反面观　G1，G2. 雄花花被片正面观　H1，H2. 雄花花被片反面观　I1，I2. 雄蕊　J. 雌花侧面观　K. 雌花花被片正面观　L. 雌花花被片反面观　M. 蒴果　N. 子房横切面　O. 雌蕊　P. 花柱和柱头

A1-A3. Habits　B1, B2. Inflorescences　C1, C2. Leaf, adaxial and abaxial view　D. Leaf, transversely sectioned　E1, E2. Staminate flower, adaxial view　F1, F2. Staminate flower, abaxial view　G1, G2. Tepals of staminate flower, adaxial view　H1, H2. Tepals of staminate flower, abaxial view　I1, I2. Androecium, lateral view　J. Pistillate flower, lateral view　K. Tepals of pistillate flower, adaxial view　L. Tepals of pistillate flower, abaxial view　M. Capsule　N. Ovary, transversely sectioned　O. Gynoecium, adaxial view　P. Styles and stigmas

铁十字秋海棠

Begonia masoniana Irmsch. ex Ziesenh.

多年生草本，具根状茎。叶片斜宽卵形至近圆形，表面浅绿色，沿主脉分布有深褐色斑纹，密被锥状长硬毛。花黄色至浅黄绿色；雄花花被片4，雌花花被片3；蒴果长圆形，被红色硬腺毛，具不等长3翅；子房1室，侧膜胎座；柱头3。

分布： 原产于中国（广西）、越南。

生境： 石灰岩斜坡、密林或灌丛下的洞穴中。

Perennial herb, rhizomatous. Leaves obliquely broadly ovate to suborbicular, adaxial surface pale green, with broad blackish brown (red abaxially) bands along main veins, with many hairy tipped pustules. Flowers yellowish to pale yellowish-green; male flowers tepals 4, female flowers tepals 3; capsule oblong, red glandular-hispid, with unequal 3 wings; ovary 1-locular, placentation parietal; styles 3.

Distribution: Native to China (Guangxi), and Vietnam.

Habitat: Rocky limestone slopes, in caves, under dense forest cover or shrubbery.

A

10cm

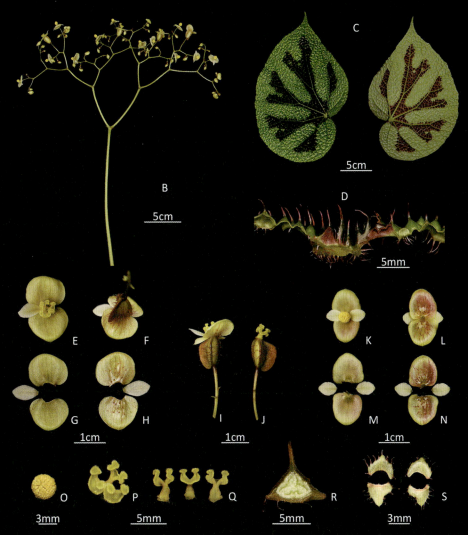

A. 植株　B. 花序　C. 叶片正反面　D. 叶片横切面　E. 雌花正面观　F. 雌花反面观
G. 雌花花被片正面观　H. 雌花花被片反面观　I. 雌花侧面观　J. 蒴果　K. 雄花正面观
L. 雄花反面观　M. 雄花花被片正面观　N. 雄花花被片反面观　O. 雄蕊　P. 雌蕊
Q. 花柱和柱头　R. 子房横切面　S. 苞片

A. Habit　B. Inflorescence　C. Leaf, adaxial and abaxial view　D. Leaf, transversely
sectioned　E. Pistillate flower, adaxial view　F. Pistillate flower, abaxial view　G. Tepals of
pistillate flower, adaxial view　H. Tepals of pistillate flower, abaxial view　I. Pistillate flower,
lateral view　J. Capsule　K. Staminate flower, adaxial view　L. Staminate flower, abaxial
view　M. Tepals of staminate flower, adaxial view　N. Tepals of staminate flower, abaxial
view　O. Androecium, adaxial view　P. Gynoecium, adaxial view　Q. Styles and stigmas
R. Ovary, transversely sectioned　S. Bracts, adaxial and abaxial view

宁明秋海棠

Begonia ningmingensis D. Fang, Y. G. Wei & C. I Peng var. *ningmingensis*

多年生草本，具根状茎，密被毛。叶片宽卵形或近圆形，前端较尖，边缘具小齿和短缘毛；表面绿色，深绿色或深褐色，沿主脉具白色斑点，背面红色。花白色或粉红色；雄花花被片4，雌花花被片3；蒴果三棱椭圆形，具不等长或近等长3翅；子房1室，侧膜胎座；花柱3。

分布： 原产于中国（广西）。

生境： 阔叶林中的石灰岩石上。

Perennial herb, rhizomatous, densely hairy. Leaves broadly ovate or suborbicular, apex acuminate, margin denticulate and ciliolate; adaxial surface green, dark green or dark brown with white maculation along major veins, abaxial surface reddish. Flowers white or pinkish; male flowers tepals 4, female flowers tepals 3; capsule trigonous-ellipsoid, with unequal or subequal 3 wings; ovary 1-locular, placentation parietal; styles 3.

Distribution: Native to China (Guangxi).

Habitat: On limestone hills in broad-leaved forests.

A

3cm

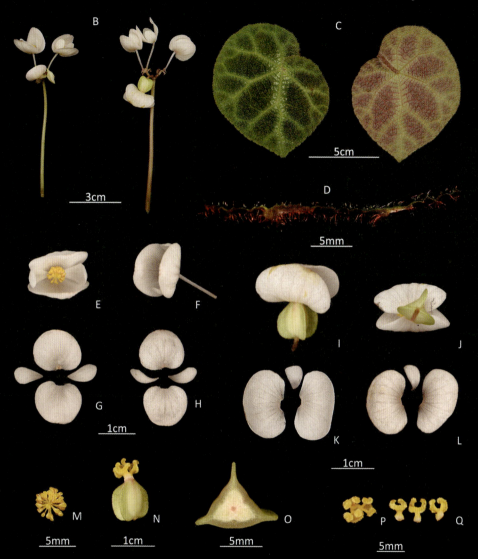

A. 植株　B. 花序　C. 叶片正反面　D. 叶片横切面　E. 雄花正面观　F. 雄花侧面观 G. 雄花花被片正面观　H. 雄花花被片反面观　I. 雌花侧面观　J. 雌花反面观　K. 雌花花被片正面观　L. 雌花花被片反面观　M. 雄蕊　N. 蒴果　O. 子房横切面　P. 雌蕊 Q. 花柱和柱头

A. Habit　B. Inflorescence　C. Leaf, adaxial and abaxial view　D. Leaf, transversely sectioned　E. Staminate flower, adaxial view　F. Staminate flower, lateral view　G. Tepals of staminate flower, adaxial view　H. Tepals of staminate flower, abaxial view　I. Pistillate flower, lateral view　J. Pistillate flower, abaxial view　K. Tepals of pistillate flower, adaxial view　L. Tepals of pistillate flower, abaxial view　M. Androecium, adaxial view　N. Capsule O. Ovary, transversely sectioned　P. Gynoecium, adaxial view　Q. Styles and stigmas

丽叶秋海棠

Begonia ningmingensis var. *bella* D. Fang, Y. G. Wei & C. I Peng

  多年生草本，根状茎。本种与宁明秋海棠非常相似，最大的区别在于该种叶片较圆，叶尖不明显；托叶全缘，无毛；果实成熟后，花被片宿存并增厚。

  分布：原产于中国（广西）。

  生境：石灰岩山上。

Perennial herb, rhizomatous. The species is very similar to *Begonia ningmingensis*, the main difference lies in the more rounded shape of its leaves and less pronounced leaf apex; the stipules are entire and without hairs; after fruiting, the tepals persist and thicken.

  Distribution: Native to China (Guangxi).

  Habitat: On limestone hills.

A

5cm

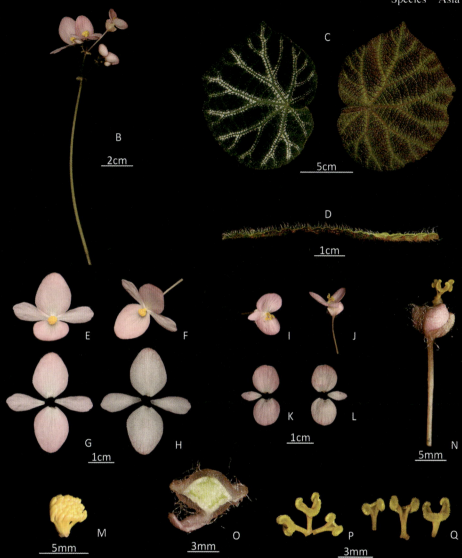

A. 植株　B. 花序　C. 叶片正反面　D. 叶片横切面　E. 雄花正面观　F. 雄花侧面观 G. 雄花花被片正面观　H. 雄花花被片反面观　I. 雌花正面观　J. 雌花侧面观　K. 雌花 花被片正面观　L. 雌花花被片反面观　M. 雄蕊　N. 蒴果　O. 子房横切面　P. 雌蕊 Q. 花柱和柱头

A. Habit　B. Inflorescence　C. Leaf, adaxial and abaxial view　D. Leaf, transversely sectioned　E. Staminate flower, adaxial view　F. Staminate flower, lateral view　G. Tepals of staminate flower, adaxial view　H. Tepals of staminate flower, abaxial view　I. Pistillate flower, adaxial view　J. Pistillate flower, lateral view　K. Tepals of pistillate flower, adaxial view　L. Tepals of pistillate flower, abaxial view　M. Androecium, lateral view　N. Capsule O. Ovary, transversely sectioned　P. Gynoecium, adaxial view　Q. Styles and stigmas

假大新秋海棠

Begonia pseudodaxinensis S. M. Ku, Yan Liu & C. I Peng

多年生草本，根状茎粗壮。叶互生，叶片宽卵形或近圆形，基部深心形，表面略具皱，边缘具小齿和短缘毛。花微香，白色；蒴果三棱椭圆形或卵状椭圆形，具不等长3翅，背翅新月形或近矩形；子房1室，侧膜胎座；柱头3。

分布： 原产于中国（广西）。

生境： 石灰岩山上。

Perennial herb, rhizomes stout. Leaves alternate, broadly ovate or suborbicular, base deeply cordate, slightly rugose, margin denticulate and ciliolate. Flowers slightly fragrant, white; capsule trigonous-ellipsoid or ovoid-ellipsoid, with unequal 3 wings, abaxial wing crescent-shaped or subrectangular; ovary 1-locular, placentation parietal; styles 3.

Distribution: Native to China (Guangxi).

Habitat: On limestone hills.

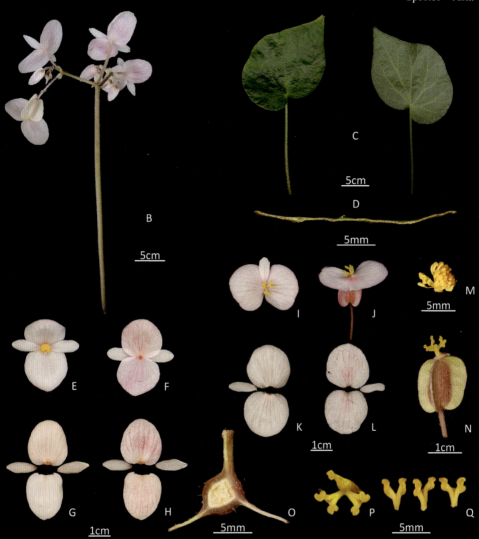

A. 植株　B. 花序　C.叶片正反面　D.叶片横切面　E.雄花正面观　F.雄花反面观 G.雄花花被片正面观　H.雄花花被片反面观　I.雌花正面观　J.雌花侧面观　K.雌花花被片正面观　L.雌花花被片反面观　M.雄蕊　N.蒴果　O.子房横切面　P.雌蕊　Q.花柱和柱头

A. Habit　B. Inflorescence　C. Leaf, adaxial and abaxial view　D. Leaf, transversely sectioned　E. Staminate flower, adaxial view　F. Staminate flower, abaxial view　G. Tepals of staminate flower, adaxial view　H. Tepals of staminate flower, abaxial view　I. Pistillate flower, adaxial view　J. Pistillate flower, lateral view　K. Tepals of pistillate flower, adaxial view　L. Tepals of pistillate flower, abaxial view　M. Androecium, lateral view　N. Capsule　O. Ovary, transversely sectioned P. Gynoecium, adaxial view　Q. Styles and stigmas

突脉秋海棠

Begonia retinervia D. Fang, D. H. Qin & C. I Peng

多年生草本，根状茎。本种与宁明秋海棠、丽叶秋海棠相似，主要区别在于其叶子较厚，纸质感强；子房呈宽三棱卵形或倒卵形，蒴果表面具卷曲的长柔毛，背翅更为突出。

分布：原产于中国（广西）。

生境：石灰岩斜坡和潮湿的石灰岩洞穴。

Perennial herb, rhizomatous. The species is similar to *Begonia ningmingensis* and *Begonia ningmingensis* var. *bella*, but distinguished primarily by its thickly chartaceous leaves; broadly trigonous-ovoid or obovoid ovary and fruit covered with crisp-villous hairs, and the much protruded abaxial wing.

Distribution: Native to China (Guangxi).

Habitat: Rocky limestone slopes and moist limestone caves.

A

5cm

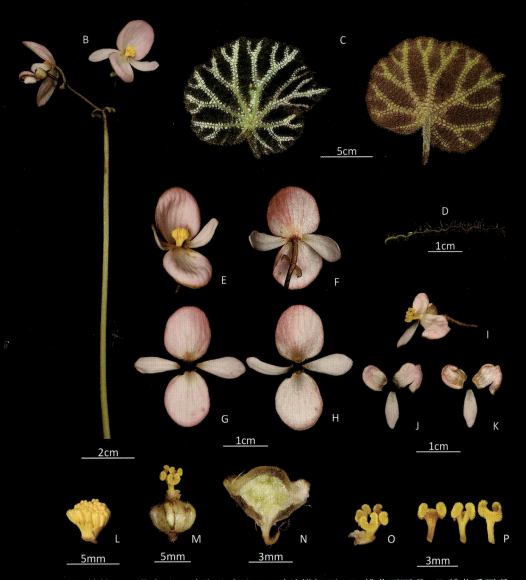

A. 植株　B. 花序　C. 叶片正反面　D. 叶片横切面　E. 雄花正面观　F. 雄花反面观
G. 雄花花被片正面观　H. 雄花花被片反面观　I. 雌花侧面观　J. 雌花花被片正面观
K. 雌花花被片反面观　L. 雄蕊　M. 蒴果　N. 子房横切面　O. 雌蕊　P. 花柱和柱头

A. Habit　B. Inflorescence　C. Leaf, adaxial and abaxial view　D. Leaf, transversely sectioned
E. Staminate flower, adaxial view　F. Staminate flower, abaxial view　G. Tepals of staminate
flower, adaxial view　H. Tepals of staminate flower, abaxial view　I. Pistillate flower,
lateral view　J. Tepals of pistillate flower, adaxial view　K. Tepals of pistillate flower,
abaxial view　L. Androecium, lateral view　M. Capsule　N. Ovary, transversely sectioned
O. Gynoecium, lateral view　P. Styles and stigmas

涩叶秋海棠

Begonia scabrifolia C. I Peng, Yan Liu & C. W. Lin

多年生草本，根状茎粗壮，匍匐。叶片宽卵形，具粗糙感；表面亮绿色、深绿至褐绿色，在主脉和侧脉间分布的银白色斑纹。花白色至粉色；蒴果宽椭圆形，粉色，具不等长3翅；子房1室，侧膜胎座；花柱3。

分布： 原产于中国（广西）和越南北部，模式标本采集于桂林植物园栽培温室。

生境： 不详。

Perennial herb, rhizomes stout, creeping. Leaves broadly ovate, surface shortly scabrous; adaxial surface bright green, dark green to brownish green, embellished with silvery white stripes between primary and secondary veins. Flowers white to pinkish; capsule widely ellipsoid, pinkish, with unequal 3 wings; ovary 1-locular, placentation parietal; styles 3.

Distribution: Native to China (Guangxi), and Northern Vietnam, type specimens pressed from plants cultivated in the Guilin Botanical Garden experimental greenhouse.

Habitat: Unknown.

A
10cm

A. 植株 B. 花序 C. 叶片正反面和叶柄 D. 叶片横切面 E. 雄花正面观 F. 雄花反面观 G. 雄花花被片正面观 H. 雄花花被片反面观 I. 雌花正面观 J. 雌花侧面观 K. 雌花花被片正面观 L. 雌花花被片反面观 M. 雄蕊 N. 蒴果 O. 子房横切面 P. 雌蕊 Q. 花柱和柱头 R. 苞片

A. Habit B. Inflorescence C. Leaf, adaxial and abaxial view, and petiole D. Leaf, transversely sectioned E. Staminate flower, adaxial view F. Staminate flower, abaxial view G. Tepals of staminate flower, adaxial view H. Tepals of staminate flower, abaxial view I. Pistillate flower, adaxial view J. Pistillate flower, lateral view K. Tepals of pistillate flower, adaxial view L. Tepals of pistillate flower, abaxial view M. Androecium, lateral view N. Capsule O. Ovary, transversely sectioned P. Gynoecium, adaxial view Q. Styles and stigmas R. Bracts, adaxial and abaxial view

半侧膜秋海棠

Begonia semiparietalis Yan Liu, S. M. Ku & C. I Peng

多年生草本，具根状茎。叶互生，宽卵形或近圆形，基部深心形，边缘具小齿和短缘毛；表面绿色、深绿色或深褐色，沿主脉分布有白色条纹，背面密被红色或浅绿色。花白色至粉色；雄花被片4，雌花花被片3；子房三棱椭圆形，具不等长或近等长3翅；子房1室，侧膜胎座，花柱3。

分布：原产于中国（广西）。

生境：阔叶林中的石灰岩石上。

Perennial herb, rhizomatous. Leaves alternate, blades broadly ovate or suborbicular, base deeply cordate, margin denticulate and ciliolate; adaxial surface green, dark green or dark brown, adorned with white maculation along major veins, abaxial surface densely reddish or pale green. Flowers white to pinkish; male flowers tepals 4, female flowers tepals 3; capsule trigonous-ellipsoid, with unequal or subequal 3 wings; ovary 1-locular, placentation parietal; styles 3.

Distribution: Native to China (Guangxi).

Habitat: On limestone hills in broad-leaved forests.

A

5cm

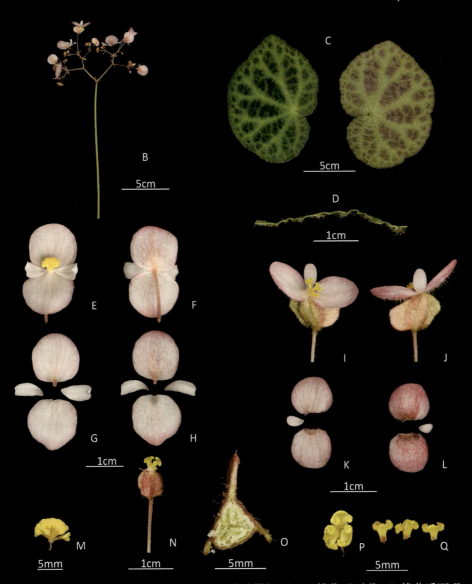

A. 植株　B. 花序　C. 叶片正反面　D. 叶片横切面　E. 雄花正面观　F. 雄花反面观
G. 雄花花被片正面观　H. 雄花花被片反面观　I, J. 雌花侧面观　K. 雌花花被片正面观
L. 雌花花被片反面观　M. 雄蕊　N. 蒴果　O. 子房横切面　P. 雌蕊　Q. 花柱和柱头

A. Habit　B. Inflorescence　C. Leaf, adaxial and abaxial view　D. Leaf, transversely
sectioned　E. Staminate flower, adaxial view　F. Staminate flower, abaxial view　G. Tepals
of staminate flower, adaxial view　H. Tepals of staminate flower, abaxial view　I, J. Pistillate
flower, lateral view　K. Tepals of pistillate flower, adaxial view　L. Tepals of pistillate flower,
abaxial view　M. Androecium, lateral view　N. Capsule　O. Ovary, transversely sectioned
P. Gynoecium, adaxial view　Q. Styles and stigmas

多花秋海棠

Begonia sinofloribunda Dorr

多年生草本，根状茎粗壮，偶有分枝，直立或铺地生长。叶盾形，叶片长圆披针形或卵状披针形，无毛，先端长渐尖或长尾状。花黄绿色；子房长圆卵形，无毛，具不等长3翅；子房3室，中轴胎座；花柱3。

分布：原产于中国（广西）。

生境：石山的荫蔽处。

Perennial herb, rhizomes stout, sometimes branched, grow upright or spread on the ground. Leaves peltate, blades oblong-lanceolate or ovate-lanceolate, glabrous, apex long acuminate or long caudate. Flowers yellowish green; capsule oblong-ovoid, glabrous, with unequal 3 wings; ovary 3-locular, placentae axile; styles 3.

Distribution: Native to China (Guangxi).

Habitat: In shady places on stony hills.

A. 植株　B. 茎　C. 叶片正反面　D. 叶片横切面　E. 雄花侧面观　F. 雄花反面观
G. 雄花花被片正面观　H. 雄花花被片反面观　I. 雌花侧面观　J. 雌花花被片正面观
K. 雌花花被片反面观　L. 蒴果　M. 雄蕊　N. 子房横切面　O. 雌蕊　P. 花柱和柱头

A. Habit　B. Stem　C. Leaf, adaxial and abaxial view　D. Leaf, transversely sectioned
E. Staminate flower, lateral view　F. Staminate flower, abaxial view　G. Tepals of staminate
flower, adaxial view　H. Tepals of staminate flower, abaxial view　I. Pistillate flower,
lateral view　J. Tepals of pistillate flower, adaxial view　K. Tepals of pistillate flower,
abaxial view　L. Capsule　M. Androecium, lateral view　N. Ovary, transversely sectioned
O. Gynoecium, lateral view　P. Styles and stigmas

近革叶秋海棠

Begonia subcoriacea C. I Peng, Yan Liu & S. M. Ku

多年生草本，根状茎粗壮。叶互生，近革质；叶片宽卵形或近圆形，基部偏斜心形；表面绿色，主脉间偶见白色斑点，背面红色。花黄绿色，较小；雄花花被片4，雌花花被片3；蒴果具不等长3翅，侧翅狭窄，背翅新月形或三角形；子房1室，侧膜胎座；花柱3。

分布： 原产于中国（广西）。

生境： 海拔250 m左右的半遮阴的石灰山坡上。

Perennial herb, rhizomes stout. Leaves alternate, subleathery; blades broadly ovate or suborbicular, base oblique cordate; adaxial surface green, with or without white maculation between major veins, abaxial surface reddish. Flowers greenish-yellow, small; male flowers tepals 4, female flowers tepals 3; capsule with 3 unequal wings, lateral wings narrower, abaxial wing crescent shaped or triangular; ovary 1-locular, placentation parietal; styles 3.

Distribution: Native to China (Guangxi).

Habitat: On semi-shaded slopes of limestone hills, *ca*. 250 m elevation.

A

10cm

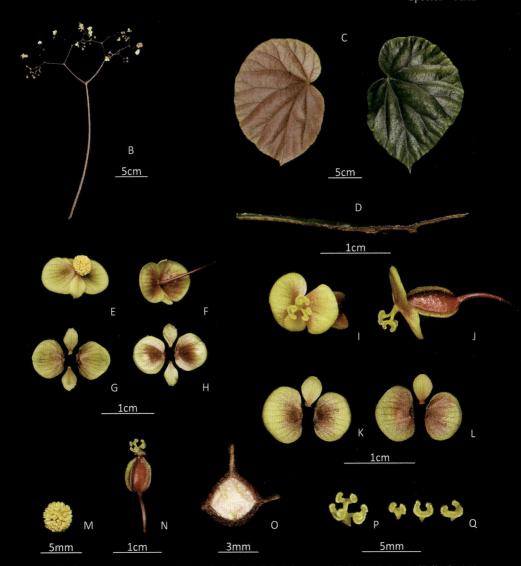

A. 植株　B. 花序　C. 叶片正反面　D. 叶片横切面　E. 雄花正面观　F. 雄花反面观 G. 雄花花被片正面观　H. 雄花花被片反面观　I. 雌花正面观　J. 雌花侧面观　K. 雌花 花被片正面观　L. 雌花花被片反面观　M. 雄蕊　N. 蒴果　O. 子房横切面　P. 雌蕊 Q. 花柱和柱头

A. Habit　B. Inflorescence　C. Leaf, adaxial and abaxial view　D. Leaf, transversely sectioned　E. Staminate flower, adaxial view　F. Staminate flower, abaxial view　G. Tepals of staminate flower, adaxial view　H. Tepals of staminate flower, abaxial view　I. Pistillate flower, adaxial view　J. Pistillate flower, lateral view　K. Tepals of pistillate flower, adaxial view　L. Tepals of pistillate flower, abaxial view　M. Androecium, adaxial view　N. Capsule O. Ovary, transversely sectioned　P. Gynoecium, adaxial view　Q. Styles and stigmas

龙虎山秋海棠

Begonia umbraculifolia Y. Wan & B. N. Chang

多年生草本，具根状茎。叶片盾状，基部有6（或7）条脉，三级脉贯穿，多数对生，凹陷，呈蜘蛛网状。花白色至粉色；雄花花被片4，雌花花被片3；蒴果长卵圆形，具不等3翅；子房1室，侧膜胎座；花柱3。

分布： 原产于中国（广西）。

生境： 海拔200～500 m的石灰岩石上、山谷或林下。

Perennial herb, rhizomatous. Leaves peltate, basally 6(or 7)-veined, tertiary veins percurrent, mostly opposite, concave, web-like. Flowers white to pinkish; male flowers tepals 4, female flowers tepals 3; capsule oblong-ovoid, with unequal 3 wings; ovary 1-locular, placentation parietal; styles 3.

Distribution: Native to China (Guangxi).

Habitat: Forest understories, valleys, on limestone, 200-500 m elevation.

A

10cm

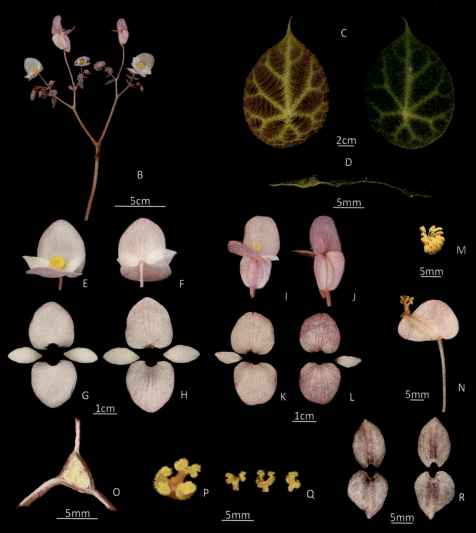

A. 植株　B. 花序　C. 叶片正反面　D. 叶片横切面　E. 雄花正面观　F. 雄花反面观 G. 雄花花被片正面观　H. 雄花花被片反面观　I. 雌花侧面观　J. 雌花反面观　K. 雌花 花被片正面观　L. 雌花花被片反面观　M. 雄蕊　N. 蒴果　O. 子房横切面　P. 雌蕊 Q. 花柱和柱头　R. 苞片

A. Habit　B. Inflorescence　C. Leaf, adaxial and abaxial view　D. Leaf, transversely sectioned　E. Staminate flower, adaxial view　F. Staminate flower, abaxial view　G. Tepals of staminate flower, adaxial view　H. Tepals of staminate flower, abaxial view　I. Pistillate flower, lateral view　J. Pistillate flower, abaxial view　K. Tepals of pistillate flower, adaxial view　L. Tepals of pistillate flower, abaxial view　M. Androecium, adaxial view　N. Capsule O. Ovary, transversely sectioned　P. Gynoecium, adaxial view　Q. Styles and stigmas R. Bracts, adaxial and abaxial view

彩纹秋海棠

Begonia variegata Y. M. Shui & W. H. Chen

多年生草本，具根状茎。叶片表面密被圆锥形的泡状隆起，沿主脉与叶缘有深褐色宽斑带。花绿色或黄绿色。该种与铁十字秋海棠相似，主要区别在于该种叶片边缘有紫色至深色的环状斑纹，花梗和花柄上有密集的腺毛。

分布：原产于越南。

生境：不详。

Perennial herb, rhizomatous. Leaf blades adaxial surface densely covered with conically bullate, with wide dark brown bands along main veins and at leaf margin. Flowers greenish or greenish-yellow. The species is similar to *Begonia masoniana*, with the main difference being that this species has a purple-darkened cyclic stripe near the margin of leaves, and dense glandular hairs on the peduncles and pedicles.

Distribution: Native to Vietnam.

Habitat: Unknown.

A

10cm

A. 植株　B. 花序　C. 叶片正反面　D. 叶片横切面　E. 雄花正面观　F. 雄花反面观 G. 雄花花被片正面观　H. 雄花花被片反面观　I. 雌花侧面观　J. 雌花正面观　K. 雌花 花被片正面观　L. 雌花花被片反面观　M. 蒴果　N. 子房横切面　O. 雄蕊　P. 雌蕊 Q. 花柱和柱头

A. Habit　B. Inflorescence　C. Leaf, adaxial and abaxial view　D. Leaf, transversely sectioned　E. Staminate flower, adaxial view　F. Staminate flower, abaxial view　G. Tepals of staminate flower, adaxial view　H. Tepals of staminate flower, abaxial view　I. Pistillate flower, lateral view　J. Pistillate flower, adaxial view　K. Tepals of pistillate flower, adaxial view　L. Tepals of pistillate flower, abaxial view　M. Capsule　N. Ovary, transversely sectioned　O. Androecium, adaxial view　P. Gynoecium, adaxial view　Q. Styles and stigmas

少瓣秋海棠

Begonia wangii T. T. Yu

多年生草本，具根状茎。叶基生，叶片卵状长圆形，先端渐尖，无毛；表面深绿色，背面红褐色。花序光滑无毛；花被片粉红色；雌雄花均为2，心形；蒴果具不等长3翅；子房3室，中轴胎座；花柱3。

分布：原产于中国（广西、云南）。

生境：灌丛中的石灰岩上。

Perennial herb, rhizomatous. Leaves basal, blades ovate-oblong, apex acuminate, glabrous; adaxial surface dark green, abaxial surface reddish brown. Inflorescences glabrous; flowers pinkish; male and female flowers tepals 2, heart-shaped; capsule with unequal 3 wings; ovary 3-locular, placentae axile; styles 3.

Distribution: Native to China (Guangxi, Yunnan).

Habitat: Limestone rocks in scrubby vegetation.

A

5cm

A. 植株　B. 花序　C. 叶片正反面和叶柄　D. 叶片横切面　E. 雄花正面观　F. 雄花反面观　G. 雄花花被片正面观　H. 雄花花被片反面观　I. 雌花正面观　J. 雌花反面观　K. 雌花花被片正面观　L. 雌花花被片反面观　M. 雄蕊　N. 蒴果　O. 子房横切面　P. 雌蕊　Q. 花柱和柱头　R. 苞片

A. Habit　B. Inflorescence　C. Leaf, adaxial and abaxial view, and petiole　D. Leaf, transversely sectioned　E. Staminate flower, adaxial view　F. Staminate flower, abaxial view　G. Tepals of staminate flower, adaxial view　H. Tepals of staminate flower, abaxial view　I. Pistillate flower, adaxial view　J. Pistillate flower, abaxial view　K. Tepals of pistillate flower, adaxial view　L. Tepals of pistillate flower, abaxial view　M. Androecium, adaxial view　N. Capsule　O. Ovary, transversely sectioned　P. Gynoecium, adaxial view　Q. Styles and stigmas　R. Bract, adaxial view

巴马秋海棠

Begonia bamaensis Yan Liu & C. I Peng

多年生草本，根状茎。叶片宽卵形或近圆形，边缘具小齿和短缘毛；表面绿色，通常在主脉间点缀有白色斑带或白色斑块，背面绿色，具皱或微皱。花白色至粉色；蒴果三棱椭圆形，具不等3翅；子房1室，侧膜胎座；花柱3。

分布： 原产于中国（广西）。

生境： 石灰岩洞穴入口处半遮阴的悬崖上。

Perennial herb, rhizomatous. Leaf blades broadly ovate or suborbicular, margin crenate denticulate and ciliolate; adaxial surface green, usually adorned with white bands or white patches between major veins, abaxial surface greenish, rugose or rugulose. Flowers white to pinkish; capsule trigonous-ellipsoid, with unequal 3 wings; ovary 1-locular, placentation parietal; styles 3.

Distribution: Native to China (Guangxi).

Habitat: On semi-shady rocky cliffs at the entrance of limestone caverns.

A

5cm

A. 植株　B. 花序　C. 叶片正反面　D. 叶片横切面　E. 雄花正面观　F. 雄花反面观 G. 雄花花被片正面观　H. 雄花花被片反面观　I. 雌花侧面观　J. 雌花反面观　K. 雌花 花被片正面观　L. 雌花花被片反面观　M. 雄蕊　N. 蒴果　O. 子房横切面　P. 雌蕊 Q. 花柱和柱头

A. Habit　B. Inflorescence　C. Leaf, adaxial and abaxial view　D. Leaf, transversely sectioned　E. Staminate flower, adaxial view　F. Staminate flower, abaxial view　G. Tepals of staminate flower, adaxial view　H. Tepals of staminate flower, abaxial view　I. Pistillate flower, lateral view　J. Pistillate flower, abaxial view　K. Tepals of pistillate flower, adaxial view　L. Tepals of pistillate flower, abaxial view　M. Androecium, adaxial view　N. Capsule O. Ovary, transversely sectioned　P. Gynoecium, adaxial view　Q. Styles and stigmas

倬云秋海棠

Begonia zhuoyuniae C. I Peng, Yan Liu & K. F. Chung

多年生草本，根状茎匍匐，纤弱。叶互生，叶片卵形到肾形，基部心形，纸质，表面略具皱。花粉色；雄花花被片4，雌花花被片3；蒴果三棱椭圆形，疏生腺状柔毛，具近等长3翅；子房1室，侧膜胎座；花柱3。

分布： 原产于中国（广西）。

生境： 石灰岩洞穴。

Perennial herb, rhizomes creeping, slender. Leaves alternate, blades ovate to reniform, base cordate, chartaceous, adaxial surface somewhat rugose. Flowers pinkish; male flowers tepals 4, female flowers tepals 3; capsule trigonous-ellipsoid, sparsely glandular-pilose, with subequal 3 wings; ovary 1-locular, placentation parietal; styles 3.

Distribution: Native to China (Guangxi).

Habitat: Limestone caves.

A. 植株　B. 花序　C. 叶片正反面　D. 叶片横切面　E. 雌花正面观　F. 雌花侧面观 G. 雌花花被片正面观　H. 雌花花被片反面观　I. 雄花正面观　J. 雄花反面观　K. 雄花花被片正面观　L. 雄花花被片反面观　M. 雄蕊　N. 蒴果　O. 子房横切面　P. 雌蕊 Q. 花柱和柱头　R. 苞片

A. Habit　B. Inflorescence　C. Leaf, adaxial and abaxial view　D. Leaf, transversely sectioned　E. Pistillate flower, adaxial view　F. Pistillate flower, lateral view　G. Tepals of pistillate flower, adaxial view　H. Tepals of pistillate flower, abaxial view　I. Staminate flower, adaxial view　J. Staminate flower, abaxial view　K. Tepals of staminate flower, adaxial view　L. Tepals of staminate flower, abaxial view　M. Androecium, lateral view N. Capsule　O. Ovary, transversely sectioned　P. Gynoecium, adaxial view　Q. Styles and stigmas　R. Bracts

靖西秋海棠

Begonia jingxiensis D. Fang & Y. G. Wei

多年生草本，具根状茎。叶片宽卵形或近圆形，叶缘密被长柔毛，幼叶尤其明显，并疏生贴伏的锈色毛，后无毛；表面亮绿色，或饰有白色或浅色马蹄形斑纹。花浅粉色或桃红色；蒴果具不等长3翅；子房3室，中轴胎座；花柱3。

分布： 原产于中国（广西）。

生境： 石灰岩山上、石灰岩洞穴入口或林下。

Perennial herb, rhizomatous. Leaf blades broadly ovate or suborbicular, margin densely villous, young leaves especially conspicuous and sparsely appressed rust-colored hairs, glabrous later; adaxial surface bright green, or adorned with a whitish or pale horseshoe-shaped maculation. Flowers pale pink or peach-red; capsule with unequal 3 wings; ovary 3-locular, placentae axile; styles 3.

Distribution: Native to China (Guangxi).

Habitat: Limestone hills and entrances to limestone caves, and forests.

A

10cm

A. 植株　B. 花序　C. 叶片正反面　D. 叶片横切面　E. 雄花正面观　F. 雄花反面观 G. 雄花花被片正面观　H. 雄花花被片反面观　I. 雌花正面观　J. 雌花侧面观　K. 雌花 花被片正面观　L. 雌花花被片反面观　M. 雄蕊　N. 蒴果　O. 子房横切面　P. 雌蕊 Q. 花柱和柱头

A. Habit　B. Inflorescence　C. Leaf, adaxial and abaxial view　D. Leaf, transversely sectioned　E. Staminate flower, adaxial view　F. Staminate flower, abaxial view　G. Tepals of staminate flower, adaxial view　H. Tepals of staminate flower, abaxial view　I. Pistillate flower, adaxial view　J. Pistillate flower, lateral view　K. Tepals of pistillate flower, adaxial view　L. Tepals of pistillate flower, abaxial view　M. Androecium, lateral view　N. Capsule O. Ovary, transversely sectioned　P. Gynoecium, adaxial view　Q. Styles and stigmas

菲律宾秋海棠组 *Begonia* Sect. *Baryandra* A. DC.

匙叶秋海棠

Begonia blancii M. Hughes & C. I Peng

多年生草本，岩生，匍匐。叶片明显肉质，倒三角匙形，边缘全缘，稍波浪状，先端宽，截形或具浅圆形裂片；表面无毛，背面叶脉有毛。花粉色，雌雄花同时开放；雌雄花被片均为4；蒴果卵形，具极不等长3翅；子房2室，中轴胎座；花柱3。

分布： 原产于菲律宾（巴拉望岛）。

生境： 原始森林荫蔽处高1 m左右的小岩石两侧。

Perennial herb, lithophytic, creeping. Leaf blades distinctly fleshy, ob-triangular spathulate, margin entire, slightly wavy, apex broad, truncate or with shallow rounded lobes; adaxial surface glabrous, abaxial surface hairy on the veins. Flowers pinkish, male and female flowers open at the same time; both male and female flowers tepals 4; capsule ovoid, with highly unequal 3 wings; ovary 2-locular, placentae axile; styles 3.

Distribution: Native to the Philippines (Palawan).

Habitat: On the sides of small (up to *ca.* 1 m high) boulders in the shade of primary forest.

A

2cm

A. 植株　B. 花序　C. 叶片正反面　D. 叶片横切面　E. 雄花正面观　F. 雄花反面观　G. 雄花花被片正面观　H. 雄花花被片反面观　I. 雌花正面观　J. 雌花侧面观　K. 雌花花被片正面观　L. 雌花花被片反面观　M. 雄蕊　N. 蒴果　O. 子房横切面　P. 雌蕊　Q. 花柱和柱头。

A. Habit　B. Inflorescence　C. Leaf, adaxial and abaxial view　D. Leaf, transversely sectioned　E. Staminate flower, adaxial view　F. Staminate flower, abaxial view　G. Tepals of staminate flower, adaxial view　H. Tepals of staminate flower, abaxial view　I. Pistillate flower, adaxial view　J. Pistillate flower, lateral view　K. Tepals of pistillate flower, adaxial view　L. Tepals of pistillate flower, abaxial view　M. Androecium, lateral view　N. Capsule O. Ovary, transversely sectioned　P. Gynoecium, lateral view　Q. Styles and stigmas

镜毅秋海棠

Begonia chingipengii Rubite

多年生草本，根状茎绿色，无毛。叶互生，叶片卵形，基部偏斜心形，先端渐尖；表面光滑有光泽，通常深绿色具明显的浅绿色叶脉，背面在绿色叶脉间呈褐红色。花深粉色或白色；蒴果浅粉色，具不等长3翅；子房3室，中轴胎座；花柱3。

分布： 原产于菲律宾（吕宋岛）。

生境： 暴露至半暴露的长满草的岩石斜坡上。

Perennial herb, rhizomes green, glabrous. Leaves alternate, blades ovate, base oblique cordate, apex acuminate; adaxial surface glabrous and glossy, generally dark green with prominent light green veins, abaxial surface maroon between the green veins. Flowers dark pink or whitish; capsule pinkish, with 3 unequal wings; ovary 3-locular, placentae axile; styles 3.

Distribution: Native to the Philippines (Luzon).

Habitat: On grassy, rocky, exposed to semi-exposed slopes.

A
5cm

A. 植株 B. 花序 C. 叶片正反面 D. 叶片横切面 E. 苞片 F. 雄花正面观
G. 雄花反面观 H. 雄花花被片正面观 I. 雄花花被片反面观 J. 雌花花被片正面观
K. 雌花花被片反面观 L. 雄蕊 M. 雌蕊 N. 花柱和柱头 O. 蒴果 P. 子房横切面

A. Habit B. Inflorescence C. Leaf, adaxial and abaxial view D. Leaf, transversely
sectioned E. Bracts, adaxial view F. Staminate flower, adaxial view G. Staminate flower,
abaxial view H. Tepals of staminate flower, adaxial view I. Tepals of staminate flower,
abaxial view J. Tepals of Pistillate flower, adaxial view K. Tepals of pistillate flower,
abaxial view L. Androecium, lateral view M. Gynoecium, adaxial view N. Styles and
stigmas O. Capsules P. Ovary, transversely sectioned

绿脉秋海棠

Begonia chloroneura P. Wilkie & Sands

多年生草本，具根状茎。叶基生，叶片卵状长圆形，先端渐尖，边缘不裂，有浅锯齿，近叶脉处及淡绿色；两面均匀覆盖着粗壮的暗红色硬毛。花白色至粉色；蒴果三棱椭圆形，具极不等长3翅；子房3室，中轴胎座；花柱3。

分布：原产于菲律宾（吕宋岛）。

生境：溪流边的岩石裂缝中。

Perennial herb, rhizomatous. Leaves basal, blades ovate-oblong, apex acuminate, margin undivided, shallowly serrate, light green near venation, both surfaces with an even covering of stout, dark red, stiff hairs. Flowers white to pinkish; capsule trigonous-ellipsoid, with highly unequal 3 wings; ovary 3-locular, placentae axile; styles 3.

Distribution: Native to the Philippines (Luzon).

Habitat: In rock crevices by streams.

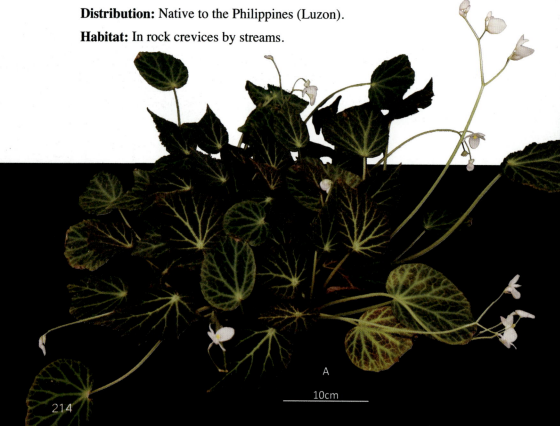

A

10cm

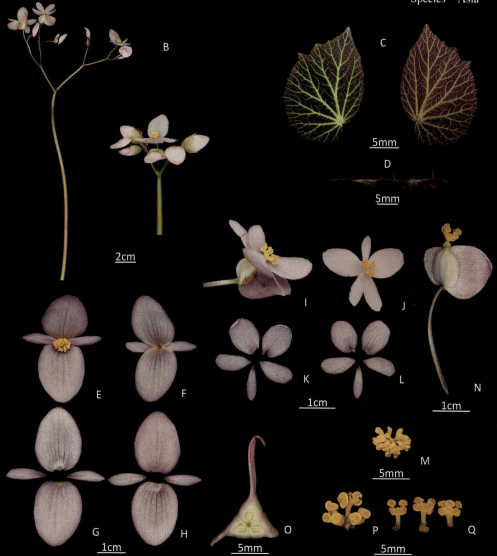

A. 植株 B. 花序 C. 叶片正反面 D. 叶片横切面 E. 雄花正面观 F. 雄花反面观 G. 雄花花被片正面观 H. 雄花花被片反面观 I. 雌花侧面观 J. 雌花正面观 K. 雌花 花被片正面观 L. 雌花花被片反面观 M. 雄蕊 N. 蒴果 O. 子房横切面 P. 雌蕊 Q. 花柱和柱头

A. Habit B. Inflorescence C. Leaf, adaxial and abaxial view D. Leaf, transversely sectioned E. Staminate flower, adaxial view F. Staminate flower, abaxial view G. Tepals of staminate flower, adaxial view H. Tepals of staminate flower, abaxial view I. Pistillate flower, lateral view J. Pistillate flower, adaxial view K. Tepals of pistillate flower, adaxial view L. Tepals of pistillate flower, abaxial view M. Androecium, adaxial view N. Capsule O. Ovary, transversely sectioned P. Gynoecium, adaxial view Q. Styles and stigmas

艳后秋海棠

Begonia cleopatrae Coyle

多年生草本，生长缓慢的岩生植物。叶基生，叶片革质，卵形至亚卵形；表面淡绿至中绿，有棕紫色花纹，沿中脉通常较浅，背面通常为红色。花白色至粉色；蒴果具极不等长3翅；子房2室，中轴胎座；柱头3。

分布： 原产于菲律宾（巴拉望）。

生境： 海拔约400 m阴湿的岩壁上。

Perennial herb, low-growing lithophyte. Leaves basal, blades coriaceous, ovate to subovate; adaxial surface pale to mid green with brownish purple patterning, usually paler along the mid vein, abaxial surface typically red. Flowers white to pinkish; capsule with highly unequal 3 wings; ovary 2-locular, placentae axile; styles 3.

Distribution: Native to the Philippines (Palawan).

Habitat: On the moist rock, *ca*. 400 m elevation.

A

5cm

A. 植株　B. 花序　C. 叶片正反面　D. 叶片横切面　E. 雄花正面观　F. 雄花反面观 G. 雄花花被片正面观　H. 雄花花被片反面观　I. 雌花侧面观　J. 雌花反面观　K. 雌花 花被片正面观　L. 雌花花被片反面观　M. 雄蕊　N. 蒴果　O. 子房横切面　P. 雌蕊 Q. 花柱和柱头

A. Habit　B. Inflorescence　C. Leaf, adaxial and abaxial view　D. Leaf, transversely sectioned　E. Staminate flower, adaxial view　F. Staminate flower, abaxial view　G. Tepals of staminate flower, adaxial view　H. Tepals of staminate flower, abaxial view　I. Pistillate flower, lateral view　J. Pistillate flower, abaxial view　K. Tepals of pistillate flower, adaxial view　L. Tepals of pistillate flower, abaxial view　M. Androecium, adaxial view　N. Capsule O. Ovary, transversely sectioned　P. Gynoecium, adaxial view　Q. Styles and stigmas

秋海棠属植物形态解剖图鉴 An Illustrated Book of Begonias

兰屿秋海棠

Begonia fenicis Merr.

多年生草本，根茎类，茎匍匐。叶片斜卵形或近圆形，膜质，先端渐尖，边缘具不规则小齿。花白色或淡粉色；雄花花被片4，雌花花被片5；蒴果卵形，具不等3翅；子房3室，中轴胎座；花柱3。

分布：原产于中国（台湾）、菲律宾（巴丹半岛）、日本（琉球群岛）。

生境：生长在海边和亚热带河谷的林地中。

Perennial herb, rhizomatous, stems creeping. Leaf blades obliquely ovate or suborbicular, membranous, apex acuminate, margin irregular with small teeth. Flowers white or pale pink; male flowers tepals 4, female flowers tepals 5; capsule ovoid, with unequal 3 wings; ovary 3-locular, placentae axile; styles 3.

Distribution: Native to China (Taiwan), the Philippines (Bataan), and Japan (Ryukyu).

Habitat: Elevated coral rocks by the sea and in woodlands in subtropical river valleys.

A

10cm

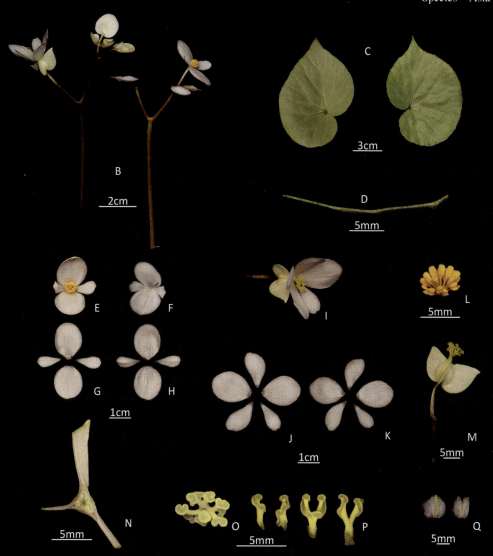

A. 植株 B. 花序 C. 叶片正反面 D. 叶片横切面 E. 雄花正面观 F. 雄花反面观
G. 雄花花被片正面观 H. 雄花花被片反面观 I. 雌花侧面观 J. 雌花花被片正面观
K. 雌花花被片反面观 L. 雄蕊 M. 蒴果 N. 子房横切面 O. 雌蕊 P. 花柱和柱头
Q. 苞片

A. Habit B. Inflorescence C. Leaf, adaxial and abaxial view D. Leaf, transversely
sectioned E. Staminate flower, adaxial view F. Staminate flower, abaxial view G. Tepals
of staminate flower, adaxial view H. Tepals of staminate flower, abaxial view I. Pistillate
flower, adaxial view J. Tepals of pistillate flower, adaxial view K. Tepals of pistillate
flower, abaxial view L. Androecium, lateral view M. Capsule N. Ovary, transversely
sectioned O. Gynoecium, adaxial view P. Styles and stigmas Q. Bracts

塔亚巴斯秋海棠
Begonia tayabensis Merr.

多年生草本，根茎类，可攀爬。叶基生，叶片盾状，卵形，先端渐尖；表面深绿色，背面红褐色，无毛。花白色至粉色；蒴果具不等长或近等长3翅；子房3室，中轴胎座，花柱3。

分布：原产于菲律宾（吕宋岛）。

生境：潮湿的林下、溪边石头上。

Perennial herb, rhizomatous, creeping. Leaves basal, blades peltate, ovate, apex acuminate; adaxial surface dark green, abaxial surface reddish brown, glabrous. Flowers white to pinkish; capsule with unequal or subequal 3 wings; ovary 3-locular, placentae axile; styles 3.

Distribution: Native to the Philippines (Luzon).

Habitat: On the forest floor or on stones along a creek.

A

4cm

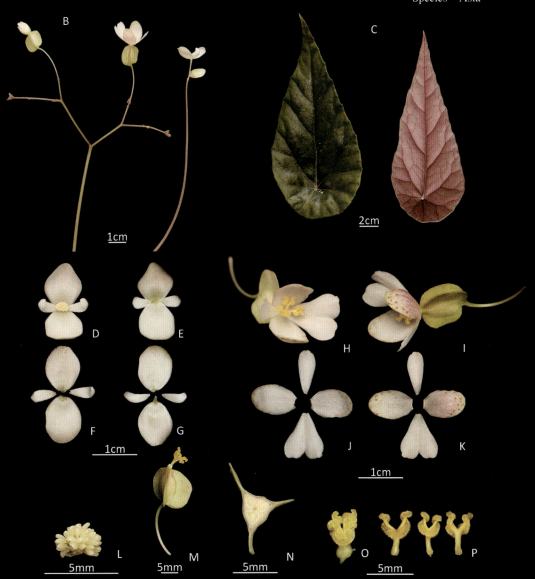

A. 植株　B. 花序　C.叶片正反面　D. 雄花正面观　E. 雄花反面观　F. 雄花花被片正面观　G. 雄花花被片反面观　H. 雌花正面观　I. 雌花侧面观　J. 雌花花被片正面观　K.雌花花被片反面观　L. 雄蕊　M. 蒴果　N. 子房横切面　O. 雌蕊　P. 花柱和柱头

A. Habit　B. Inflorescence　C. Leaf, adaxial and abaxial view　D. Staminate flower, adaxial view　E. Staminate flower, abaxial view　F. Tepals of staminate flower, adaxial view G. Tepals of staminate flower, abaxial view　H. Pistillate flower, adaxial view　I. Pistillate flower, lateral view　J. Tepals of pistillate flower, adaxial view　K. Tepals of pistillate flower, abaxial view　L. Androecium, lateral view　M. Capsule　N. Ovary, transversely sectioned O. Gynoecium, lateral view　P. Styles and stigmas

根茎单座组 *Begonia* Sect. *Jackia* M. Hughes

蛛丝秋海棠

Begonia araneumoides Ardi & Girm.

多年生草本，根茎类，匍匐，在节上生根。叶互生，叶片卵形；表面主脉呈白色，微皱，三级脉排列如蜘蛛网。花白色；蒴果椭圆形，具等长3翅；子房3室，中轴胎座；柱头3。

分布：原产于印度尼西亚（苏门答腊）。

生境：低地森林里。

Perennial herb, rhizomatous, creeping, rooting at nodes. Leaves alternate, blades ovate; adaxial surface variegated with white primary veins, rugulose, tertiary veins arranged like a spider's web. Flowers white; capsule ellipsoid, with equal 3 wings; ovary 3-locular, placentae axile; styles 3.

Distribution: Native to Indonesia (Sumatra).

Habitat: In lowland forest.

A

3cm

A. 植株　B. 花序　C. 叶片正反面　D. 叶片横切面　E. 雄花正面观　F. 雄花侧面观 G. 雄花花被片正面观　H. 雄花花被片反面观　I. 雌花正面观　J. 雌花侧面观　K. 雌花 花被片正面观　L. 雌花花被片反面观　M. 雄蕊　N. 蒴果　O. 子房横切面　P. 雌蕊 Q. 花柱和柱头

A. Habit　B. Inflorescence　C. Leaf, adaxial and abaxial view　D. Leaf, transversely sectioned　E. Staminate flower, adaxial view　F. Staminate flower, lateral view　G. Tepals of staminate flower, adaxial view　H. Tepals of staminate flower, abaxial view　I. Pistillate flower, adaxial view　J. Pistillate flower, lateral view　K. Tepals of pistillate flower, adaxial view　L. Tepals of pistillate flower, abaxial view　M. Androecium, lateral view　N. Capsule O. Ovary, transversely sectioned　P. Gynoecium, adaxial view　Q. Styles and stigmas

革叶秋海棠

Begonia coriacea Hassk.

多年生草本，根茎类。叶柄长，叶片呈盾状，近圆形，光滑无毛，边缘具圆锯齿；表面呈橄榄色，背面淡红色。花粉红色，开花先雄后雌；蒴果无毛，具近等长3翅；子房3室，中轴胎座；花柱3。

分布： 原产于印度尼西亚（爪哇、苏门答腊、巴厘）。

生境： 石灰岩壁上。

Perennial herb, rhizomatous. Petioles long, leaf blades peltate, suborbicular, glabrous, margin minutely crenate; adaxial surface olive, abaxial surface light reddish. Flowers pinkish, male flowers opening before female flowers; capsule hairless, with subequal 3 wings; ovary 3-locular, placentae axial; styles 3.

Distribution: Native to Indonesia (Java, Sumatra, Bali).

Habitat: In limestone rocky.

A

5cm

A. 植株　B. 花序　C. 叶片正反面　D. 雄花正面观　E. 雄花反面观　F. 雄花花被片正面观　G. 雄花花被片反面观　H. 雌花正面观　I. 雌花侧面观　J. 雌花花被片正面观　K. 雌花花被片反面观　L. 雄蕊　M. 蒴果　N. 子房横切面　O. 雌蕊　P. 花柱和柱头

A. Habit　B. Inflorescence　C. Leaf, adaxial and abaxial view　D. Staminate flower, adaxial view　E. Staminate flower, abaxial view　F. Tepals of staminate flower, adaxial view　G. Tepals of staminate flower, abaxial view　H. Pistillate flower, adaxial view　I. Pistillate flower, lateral view　J. Tepals of pistillate flower, adaxial view　K. Tepals of pistillate flower, abaxial view　L. Androecium, adaxial view　M. Capsule　N. Ovary, transversely sectioned　O. Gynoecium, lateral view　P. Styles and stigmas

纳土纳秋海棠

Begonia natunaensis C. W. Lin & C. I Peng

多年生草本，根状茎浅绿色或带红色。叶片盾状，近圆形；表面浅绿色，具皱，无毛，背面脉间呈浅色或粉红色。花粉色或白色；雄花较小；蒴果三棱圆形，具近等长3翅；子房3室，中轴胎座；花柱3。

分布： 原产于印度尼西亚（纳土纳岛）。

生境： 海拔约100 m的瀑布地区潮湿的砂岩悬崖上。

Perennial herb, rhizomes light green or reddish. Leaves peltate, suborbicular; adaxial surface light green, rugose, glabrous, abaxial surface pale or pinkish between venation. Flowers pink or white; male flowers small; capsule trigonous-orbicular, with subequal wings; ovary 3-locular, placentae axile; styles 3.

Distribution: Native to Indonesia (Natuna Island).

Habitat: On wet sandstone cliffs in a waterfall area, *ca.* 100 m elevation.

A

20cm

A. 植株　B. 花序　C. 叶片正反面　D. 叶片横切面　E. 雌花正面观　F. 雌花侧面观
G. 雌花花被片正面观　H. 雌花花被片反面观　I. 雄花侧面观　J. 雄花花被片正面观
K. 雄花花被片反面观　L. 雄蕊　M. 蒴果　N. 子房横切面　O. 雌蕊　P. 花柱和柱头
Q. 苞片

A. Habit　B. Inflorescence　C. Leaf, adaxial and abaxial view　D. Leaf, transversely
sectioned　E. Pistillate flower, adaxial view　F. Pistillate flower, lateral view　G. Tepals
of pistillate flower, adaxial view　H. Tepals of pistillate flower, abaxial view　I. Staminate
flower, lateral view　J. Tepals of staminate flower, adaxial view　K. Tepals of staminate
flower, abaxial view　L. Androecium, lateral view　M. Capsule　N. Ovary, transversely
sectioned　O. Gynoecium, adaxial view　P. Styles and stigmas　Q. Bracts, adaxial view and
abaxial view

等翅组 *Begonia* Sect. *Petermannia* (Klotzsch) A. DC.

龙胄秋海棠
Begonia dracopelta Ardi

多年生草本，株高约20 cm。叶互生，椭圆形，基部心形，先端渐尖，边缘具小齿；表面光滑无毛，深紫色发亮，叶脉淡绿色凹陷，背面颜色较浅，脉上被毛。花白色至粉色；蒴果卵圆形，粉色，具等长3翅；子房3室，中轴胎座；花柱3。

分布： 原产于印度尼西亚（西加里曼丹）、马来西亚（沙捞越）。

生境： 瀑布旁边潮湿的岩石上。

Perennial herb, up to *ca*. 20 cm tall. Leaves alternate, elliptic, base cordate and apex acuminate, margin denticulate; adaxial surface glabrous, shiny dark purple, veins pale green depressed, abaxial surface paler, with hairs on the veins. Flowers white to pinkish; capsule ovoid, pinkish, with subequal 3 wings; ovary 3-locular, placentae axile; styles 3.

Distribution: Native to Indonesia (the west Kalimantan), and Malaysia (Sarawak).

Habitat: On vertical moist rocks in waterfalls.

A

10cm

A. 植株　B. 花序　C. 叶片正反面　D. 叶片横切面　E. 雄花正面观　F. 雄花反面观 G. 雄花花被片正面观　H. 雄花花被片反面观　I. 雌花侧面观　J. 雌花正面观　K. 雌花 花被片正面观　L. 雌花花被片反面观　M. 雄蕊　N. 蒴果　O. 子房横切面　P. 雌蕊 Q. 花柱和柱头

A. Habit　B. Inflorescence　C. Leaf, adaxial and abaxial view　D. Leaf, transversely sectioned　E. Staminate flower, adaxial view　F. Staminate flower, abaxial view　G. Tepals of staminate flower, adaxial view　H. Tepals of staminate flower, abaxial view　I. Pistillate flower, lateral view　J. Pistillate flower, adaxial view　K. Tepals of pistillate flower, adaxial view　L. Tepals of pistillate flower, abaxial view　M. Androecium, lateral view　N. Capsule O. Ovary, transversely sectioned　P. Gynoecium, lateral view　Q. Styles and stigmas

柔浩秋海棠

Begonia jiewhoei Kiew

多年生草本，竹节状，节膨大。叶片宽卵形；表面天鹅绒质感，孔雀石般绿色，主脉间有大片银灰色斑点，背面通常为红色。总状花序，雌花先开；蒴果三棱椭圆形，浅绿色，具不等长3翅；子房3室，中轴胎座；花柱3。

分布：原产于马来西亚（马来半岛）。

生境：岩石裂缝或悬崖底部的石灰石土壤上。

Perennial herb, cane-like, nodes swollen. Leaf blades broadly ovate; adaxial surface velvety, deeply malachite green with large grey-silver spots between the main veins, abaxial usually reddish. Inflorescences racemose, female flowers open first; capsule trigonous-ellipsoid, light green, with equal or unequal 3 wings; ovary 3-locular, placentae axile; styles 3.

Distribution: Native to Malaysia (Peninsular Malaysia).

Habitat: In rock fissures or on limestone-derived soil at the base of cliffs.

A

5cm

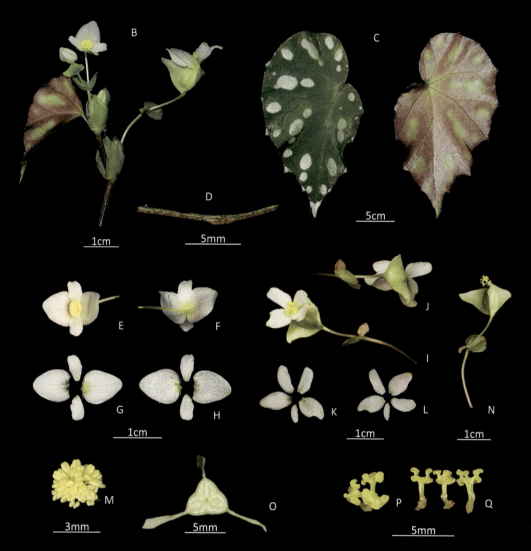

A. 植株　B. 花序　C. 叶片正反面　D. 叶片横切面　E. 雄花正面观　F. 雄花反面观
G. 雄花花被片正面观　H. 雄花花被片反面观　I. 雌花侧面观　J. 雌花反面观　K. 雌花
花被片正面观　L. 雌花花被片反面观　M. 雄蕊　N. 蒴果　O. 子房横切面　P. 雌蕊
Q. 花柱和柱头

A. Habit　B. Inflorescence　C. Leaf, adaxial and abaxial view　D. Leaf, transversely
sectioned　E. Staminate flower, adaxial view　F. Staminate flower, abaxial view　G. Tepals
of staminate flower, adaxial view　H. Tepals of staminate flower, abaxial view　I. Pistillate
flower, lateral view　J. Pistillate flower, abaxial view　K. Tepals of pistillate flower, adaxial
view　L. Tepals of pistillate flower, abaxial view　M. Androecium, adaxial view　N. Capsule
O. Ovary, transversely sectioned　P. Gynoecium, adaxial view　Q. Styles and stigmas

波利略秋海棠

Begonia polilloensis Tebbitt

多年生草本，茎直立，高约30 cm。叶互生，单叶，叶片掌状全裂，最大的裂片二回羽状分裂，其他裂片羽状浅裂至深裂，末回裂片具牙齿、先端具芒尖。花粉色，花被片基部颜色较深；蒴果椭圆形至窄倒卵形，具不等或近等长3翅；子房3室，中轴胎座；花柱3。

分布：原产于菲律宾（吕宋岛、内格罗斯岛、波利略）。

生境：不详。

Perennial herb, stems erect, up to *ca*. 30 cm tall. Leaves alternate, simple, and palmately divided, with the largest lobes being bipinnately divided and the other lobes being pinnately shallow to deeply divided, with toothed margins and awned tips. Flowers pink, deep at base of tepals; capsule elliptical to narrowly obovate, with equal or subequal 3 wings; ovary 3-locular, placentae axile; styles 3.

Distribution: Native to the Philippines (Luzon, Negros, Polillo).

Habitat: Unknown.

A. 植株　B. 花序　C. 叶片正反面　D. 叶片横切面　E. 雄花正面观　F. 雄花反面观 G. 雄花花被片正面观　H. 雄花花被片反面观　I. 雌花正面观　J. 雌花侧面观　K. 雌花 花被片正面观　L. 雌花花被片反面观　M. 雄蕊　N. 蒴果　O. 子房横切面　P. 雌蕊 Q. 花柱和柱头

A. Habit　B. Inflorescence　C. Leaf, adaxial and abaxial view　D. Leaf, transversely sectioned　E. Staminate flower, adaxial view　F. Staminate flower, abaxial view　G. Tepals of staminate flower, adaxial view　H. Tepals of staminate flower, abaxial view　I. Pistillate flower, adaxial view　J. Pistillate flower, lateral view　K. Tepals of pistillate flower, adaxial view　L. Tepals of pistillate flower, abaxial view　M. Androecium, lateral view　N. Capsule O. Ovary, transversely sectioned　P. Gynoecium, adaxial view　Q. Styles and stigmas

保亭秋海棠

Begonia sublongipes Y. M. Shui

多年生草本，具匍匐茎，茎疏生糙硬毛。叶互生，叶片卵状椭圆形，边缘疏生浅锯齿；表面无毛，背面叶脉上有短柔毛。花白色至粉色；蒴果宽倒三角形，无毛，具近等长3翅；子房3室，中轴胎座；花柱3。

分布：原产于中国（海南）。

生境：林下阴湿山谷或溪流旁石壁上。

Perennial herb, prostrate, stems sparsely hispid. Leaves stem-borne, blades ovate-elliptic, margin sparsely serrate; adaxial surface glabrous, abaxial surface pubescent on veins. Flowers white to pinkish; capsule broadly inverted triangular, glabrous, with subequal 3 wings; ovary 3-locular, placentae axile; styles 3.

Distribution: Native to China (Hainan).

Habitat: In shady and humid valleys under the forest or on rocky walls near streams.

A

20cm

A. 植株　B. 茎　C. 花序　D. 叶片正反面　E. 雄花正面观　F. 雄花反面观　G. 雄花花被片正面观　H. 雄花花被片反面观　I. 雌花侧面观　J. 雌花反面观　K. 雌花花被片正面观　L. 雌花花被片反面观　M. 雄蕊　N. 蒴果　O. 子房横切面　P. 雌蕊　Q. 花柱和柱头　P. 雌蕊　Q. 花柱和柱头

A. Habit　B. Stem　C. Inflorescence　D. Leaf, adaxial and abaxial view　E. Staminate flower, adaxial view　F. Staminate flower, abaxial view　G. Tepals of staminate flower, adaxial view　H. Tepals of staminate flower, abaxial view　I. Pistillate flower, lateral view J. Pistillate flower, abaxial view　K. Tepals of pistillate flower, adaxial view　L. Tepals of pistillate flower, abaxial view　M. Androecium, adaxial view　N. Capsule　O. Ovary, transversely sectioned　P. Gynoecium, adaxial view　Q. Styles and stigmas　P. Gynoecium, adaxial view　Q. Styles and stigmas

肿节组
Begonia Sect. Boisiana Y. M. Shui, H-Y. Chen & W. K. Dong

布瓦秋海棠
Begonia boisiana Gagnep.

多年生草本，丛生类，茎直立。叶片呈狭卵形，革质；表面零星分布有腺点，背面红色叶脉明显。总状花序；花白色；雄花花被片4，雌花花被片5；蒴果粉红色，具不等长3翅；子房3室，中轴胎座；花柱3。

分布：原产于越南。

生境：湿热的环境。

Perennial herb, shrub-like, stems upright. Leaf blades narrowly ovate, leathery; adaxial surface with scattered glandular dots, abaxial surface with obviously red veins. Inflorescences racemose; flowers white; male flowers tepals 4, female flowers tepals 5; capsule pinkish, with unequal 3 wings; ovary 3-locular, placentae axile; styles 3.

Distribution: Native to Vietnam.

Habitat: In wet tropical biome.

A

10cm

A. 植株　B. 花序　C. 叶片正反面　D. 叶片横切　E. 雄花正面观　F. 雄花反面观
G. 雄花花被片正面观　H. 雄花花被片反面观　I. 雌花正面观　J. 雌花侧面观　K. 雌花
花被片正面观　L. 雌花花被片反面观　M. 雄蕊　N. 蒴果　O. 子房横切面　P. 雌蕊
Q. 花柱和柱头

A. Habit　B. Inflorescence　C. Leaf, adaxial and abaxial view　D. Leaf, transversely
sectioned　E. Staminate flower, adaxial view　F. Staminate flower, abaxial view　G. Tepals
of staminate flower, adaxial view　H. Tepals of staminate flower, abaxial view　I. Pistillate
flower, adaxial view　J. Pistillate flower, lateral view　K. Tepals of pistillate flower, adaxial
view　L. Tepals of pistillate flower, abaxial view　M. Androecium, lateral view　N. Capsule
O. Ovary, transversely sectioned　P. Gynoecium, adaxial view　Q. Styles and stigmas

海南秋海棠组
Begonia Sect. *Hainania* Y. M. Shui & W. H. Chen

盾叶秋海棠
Begonia peltatifolia Li

多年生草本，耐旱。叶片盾状，近圆形，表面光滑无毛。聚伞状花序，花梗光滑；雄花花被片4，雌花花被片2，近无毛；蒴果倒卵形，无毛，具不等长3翅；子房3室，中轴胎座；花柱3。

分布： 原产于中国（海南）。

生境： 干旱的岩壁上。

Perennial herb, drought-tolerant. Leaf blades peltate, suborbicular, glabrous. Inflorescences racemose, peduncles glabrous; male flowers tepals 4, female flowers tepals 2, glabrous; capsule obovate, glabrous, with unequal 3 wings; ovary 3-locular, placentae axile; styles 3.

Distribution: Native to China (Hainan).

Habitat: On dry rock walls.

A

5cm

A. 植株　B. 花序　C. 叶片正反面　D. 叶片横切面　E. 雄花正面观　F. 雄花反面观 G. 雄花花被片正面观　H. 雄花花被片反面观　I. 雌花正面观　J. 雌花侧面观　K. 雌花花被片正面观　L. 雌花花被片反面观　M. 雄蕊　N. 蒴果　O. 子房横切面　P. 雌蕊 Q. 花柱和柱头

A. Habit　B. Inflorescence　C. Leaf, adaxial and abaxial view　D. Leaf, transversely sectioned　E. Staminate flower, adaxial view　F. Staminate flower, abaxial view　G. Tepals of staminate flower, adaxial view　H. Tepals of staminate flower, abaxial view　I. Pistillate flower, adaxial view　J. Pistillate flower, lateral view　K. Tepals of pistillate flower, adaxial view　L. Tepals of pistillate flower, abaxial view　M. Androecium, adaxial view　N. Capsule O. Ovary, transversely sectioned　P. Gynoecium, adaxial view　Q. Styles and stigmas

克娄巴特拉秋海棠
Begonia 'Cleopatra'

秋海棠品种 Cultivars

橙红秋海棠
Begonia 'Orange Rubra'

　　该品种由美国育种者莱斯利·伍德里夫（Leslie Woodriff）于1947年杂交育成，母本为双色秋海棠（*Begonia dichroa*），父本为珊瑚红秋海棠（*Begonia* 'Coral Rubra'）。

　　该品种叶片淡绿色，带有少量银色斑点，花橙色，在中国广东省深圳市可常年开花。

The cultivar was hybridized by American breeder Leslie Woodriff in 1947, with the female parent being *Begonia dichroa* and the male parent being *Begonia* 'Coral Rubra'. Its leaves are light green with a few silver spots and orange flowers. It can bloom throughout the year in Shenzhen, Guangdong, China.

A

10cm

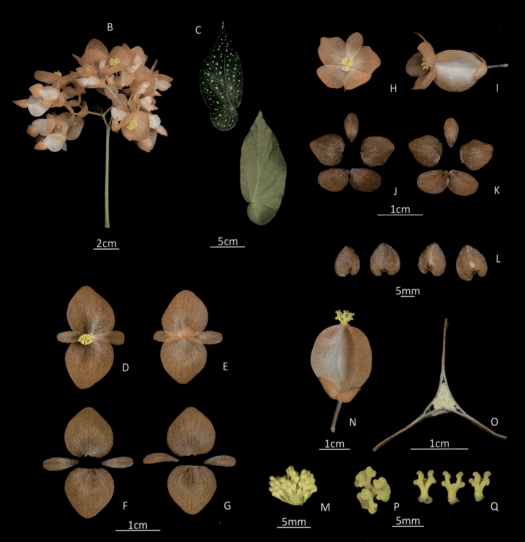

A. 植株　B. 花序　C. 叶片正反面　D. 雄花正面观　E. 雄花反面观　F. 雄花花被片正面观　G. 雄花花被片反面观　H. 雌花正面观　I. 雌花侧面观　J. 雌花花被片正面观　K. 雌花花被片反面观　L. 小苞片　M. 雄蕊　N. 蒴果　O. 子房横切面　P. 雌蕊　Q. 花柱和柱头

A. Habit　B. Inflorescence　C. Leaf, adaxial and abaxial view　D. Staminate flower, adaxial view　E. Staminate flower, abaxial view　F. Tepals of staminate flower, adaxial view　G. Tepals of staminate flower, abaxial view　H. Pistillate flower, adaxial view　I. Pistillate flower, lateral view　J. Tepals of pistillate flower, adaxial view　K. Tepals of pistillate flower, abaxial view　L. Bracteoles　M. Androecium, lateral view　N. Capsule　O. Ovary, transversely sectioned　P. Gynoecium, adaxial view　Q. Styles and stigmas

珊瑚秋海棠

Begonia 'Pinafore'

该品种中文别名大红竹节秋海棠，由美国育种者欧内斯特·马丁（Ernest E. Martin）于1951年育成，母本为伊莲秋海棠（*Begonia* 'Elaine'），父本未知。

该品种叶片表面墨绿色，背面深红色，花深粉色，量大，全年可陆续开花，具有很高的观赏价值。

The cultivar was created by American breeder Ernest E. Martin in 1951, with the female parent being *Begonia* 'Elaine' and the male parent being unknown. The foliage of this cultivar is dark green on the adaxial surface and deep red on the abaxial surface. Its flowers are deep pink, and it blooms continuously throughout the year, making it highly ornamental and visually appealing.

A

10cm

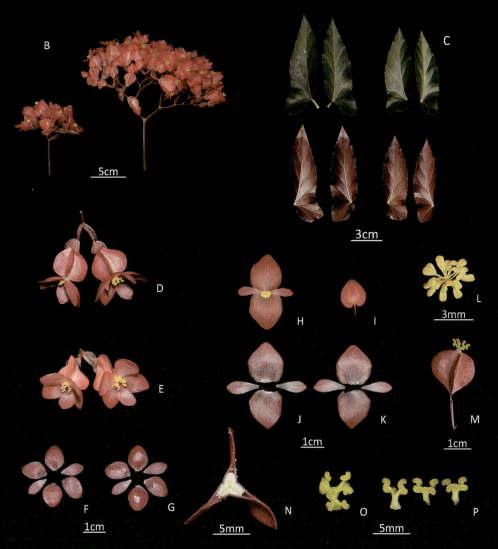

A. 植株　B. 花序　C. 叶片正反面　D. 雌花侧面观　E. 雌花正面观　F. 雌花花被片正面观　G. 雌花花被片反面观　H. 雄花正面观　I. 未开放雄花侧面观　J. 雄花花被片正面观　K. 雄花花被片反面观　L. 雄蕊　M. 蒴果　N. 子房横切面　O. 雌蕊　P. 花柱和柱头

A. Habit　B. Inflorescence　C. Leaf, adaxial and abaxial view　D. Pistillate flower, lateral view　E. Pistillate flower, adaxial view　F. Tepals of pistillate flower, adaxial view G. Tepals of pistillate flower, abaxial view　H. Staminate flower, adaxial view　I. Unopened staminate flower, lateral view　J. Tepals of staminate flower, adaxial view　K. Tepals of staminate flower, abaxial view　L. Androecium, adaxial view　M. Capsule　N. Ovary, transversely sectioned　O. Gynoecium, adaxial view　P. Styles and stigmas

布隆泽秋海棠

Begonia 'Bronze King'

该品种由美国育种者于 1931 年育成，亲本信息不详。

该品种叶片长卵形，掌状浅裂至中裂，叶基部螺旋状卷曲，表面墨绿色，具有少量银色斑点，花大，粉色。

The cultivar was developed by an American breeder in 1931, but the parentage information is unknown. The leaf blades are oblong-ovate with shallow to moderately divided lobes in a palmate arrangement, base spirally curled, adaxial surface dark green with a few silver spots, and the flowers are large and pink.

A

5cm

A. 植株　B. 花序　C. 叶片正反面　D. 雄花正面观　E. 雄花反面观　F. 雄花花被片正面观　G. 雄花花被片反面观　H. 雌花正面观　I. 雌花反面观　J. 雌花花被片正面观　K. 雌花花被片反面观　L. 雄蕊　M. 蒴果　N. 子房横切面　O. 雌蕊　P. 花柱和柱头

A. Habit　B. Inflorescence　C. Leaf, adaxial and abaxial view　D. Staminate flower, adaxial view　E. Staminate flower, abaxial view　F. Tepals of staminate flower, adaxial view　G. Tepals of staminate flower, abaxial view　H. Pistillate flower, adaxial view　I. Pistillate flower, abaxial view　J. Tepals of pistillate flower, adaxial view　K. Tepals of pistillate flower, abaxial view　L. Androecium, adaxial view　M. Capsule　N. Ovary, transversely sectioned　O. Gynoecium, adaxial view　P. Styles and stigmas

蜗牛秋海棠

Begonia 'Escargot'

该品种最早在荷兰培育，2000年由麦哈钦森（McHutchinson）引入美国。

叶基螺旋状卷曲，并继承了亲本大王秋海棠的银白色环状斑纹，叶缘和中心深绿色带紫红色。

The cultivar was originally bred in the Netherlands and was introduced to the United States by McHutchinson in 2000. The leaves have a spiral curl at the base and inherit the silver-white circular markings from the parent *Begonia rex*. The leaf margins and center are deep green with a purplish-red hue.

A

5cm

A. 植株　B. 花序　C. 苞片　D. 叶片正反面　E. 雄花正面观　F. 雄花反面观
G. 雄花花被片正面观　H. 雄花花被片反面观　I. 雌花正面观　J. 雌花反面观　K. 雌花
花被片正面观　L. 雌花花被片反面观　M. 雄蕊　N. 蒴果　O. 子房横切面　P. 雌蕊
Q. 花柱和柱头

A. Habit　B. Inflorescence　C. Bracts　D. Leaf, adaxial and abaxial view　E. Staminate
flower, adaxial view　F. Staminate flower, abaxial view　G. Tepals of staminate flower,
adaxial view　H. Tepals of staminate flower, abaxial view　I. Pistillate flower, adaxial view
J. Pistillate flower, abaxial view　K. Tepals of pistillate flower, adaxial view　L. Tepals
of pistillate flower, abaxial view　M. Androecium, lateral view　N. Capsule　O. Ovary,
transversely sectioned　P. Gynoecium, adaxial view　Q. Styles and stigmas

费多尔秋海棠

Begonia 'Fedor'

　　该品种由荷兰育种者安东·胡夫纳格尔斯（Antoon Hoefnagels）以火山秋海棠（*Begonia* 'Volcano'）为母本，*Begonia* U288 为父本杂交选育而来，并于2005年发布。

　　该品种叶片表面掌状叶脉点缀银白色斑纹，花大，花被片粉色。

The cultivar was hybridized and selected by Dutch breeder Antoon Hoefnagels, using *Begonia* 'Volcano' as the female parent and *Begonia* U288 as the male parent, and it was made public in 2005. The leaves are adorned with silver-white markings along the palmate veins on the adaxial surface, and its tepals are large, and pink.

A

10cm

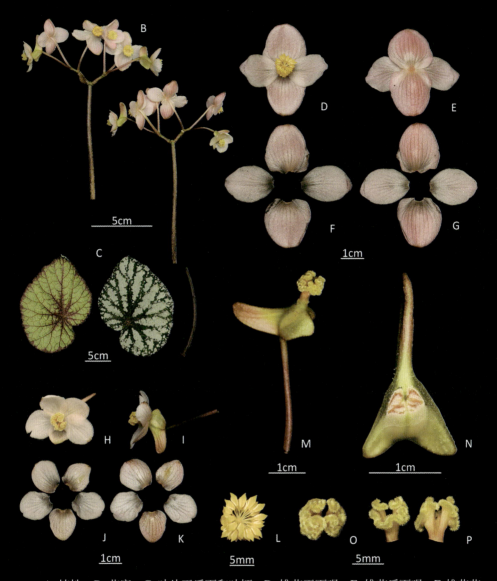

A. 植株　B. 花序　C. 叶片正反面和叶柄　D. 雄花正面观　E. 雄花反面观　F. 雄花花被片正面观　G. 雄花花被片反面观　H. 雌花正面观　I. 雌花侧面观　J. 雌花花被片正面观 K. 雌花花被片反面观　L. 雄蕊　M. 蒴果　N. 子房横切面　O. 雌蕊　P. 花柱和柱头

A. Habit　B. Inflorescence　C. Leaf, adaxial and abaxial view, and petiole D. Staminate flower, adaxial view　E. Staminate flower, abaxial view　F. Tepals of staminate flower, adaxial view　G. Tepals of staminate flower, abaxial view　H. Pistillate flower, adaxial view　I. Pistillate flower, lateral view　J. Tepals of pistillate flower, adaxial view K. Tepals of pistillate flower, abaxial view　L. Androecium, adaxial view　M. Capsule　N. Ovary, transversely sectioned　O. Gynoecium, adaxial view　P. Styles and stigmas

海伦刘易斯秋海棠

Begonia 'Helen Lewis'

　　该品种由美国育种者海伦·刘易斯（Helen Lewis）于1940年育成。

　　该品种叶片具有天鹅绒般光泽，黑紫色带银白色环状斑纹，幼叶环状斑纹上有粉紫色晕染。

The cultivar was hybridized by American breeder Helen Lewis in 1940. It features leaves with a velvety sheen, black-purple color with silver-white circular spots, and the young leaves have a pink-purple tinge on the circular spots.

A

5cm

A. 植株　B. 花序　C. 叶片正反面，叶柄　D. 雄花正面观　E. 雄花反面观　F. 雄花花被片正面观　G. 雄花花被片反面观　H. 雌花正面观　I. 雌花反面观　J. 雌花花被片正面观　K. 雌花花被片反面观　L. 雄蕊　M. 雌蕊　N. 花柱和柱头　O. 蒴果　P. 子房横切面

A. Habit　B. Inflorescence　C. Leaf, adaxial and abaxial view, petiole　D. Staminate flower, adaxial view　E. Staminate flower, abaxial view　F. Tepals of staminate flower, adaxial view　G. Tepals of staminate flower, abaxial view　H. Pistillate flower, adaxial view I. Pistillate flower, abaxial view　J. Tepals of pistillate flower, adaxial view　K. Tepals of pistillate flower, abaxial view　L. Androecium, lateral view　M. Gynoecium, adaxial view N. Styles and stigmas　O. Capsule　P. Ovary, transversely sectioned

蓝色树莓秋海棠

Begonia 'Jurassic Jr. Purple Spec'

该品种为美国Nursery公司开发的侏罗纪（Jurassic）系列的品种之一。

该品种叶片灰绿色，具金属光泽，叶缘和基部呈蓝紫色，叶基微螺旋状，叶缘不规则皱波状。

The cultivar is one of the 'Jurassic' series developed by the Nursery company. This particular cultivar boasts mesmerizing grey-green leaves that gleam with a metallic sheen. Its leaf margin and base are adorned with a captivating blue-purple hue, adding a touch of elegance to its appearance. Notably, the leaf base showcases a delicate spiral pattern, while the leaf edges exhibit charmingly irregular and wavy ripples, enhancing its allure.

A

10cm

A. 植株　B. 花序　C. 苞片　D. 叶片正反面　E. 雄花正面观　F. 雄花反面观 G. 雄花花被片正面观　H. 雄花花被片反面观　I. 雌花正面观　J. 雌花侧面观　K. 雌花 花被片正面观　L. 雌花花被片反面观　M. 雄蕊　N. 蒴果　O. 子房横切面　P. 雌蕊 Q. 花柱和柱头

A. Habit　B. Inflorescence　C. Bracts　D. Leaf, adaxial and abaxial view　E. Staminate flower, adaxial view　F. Staminate flower, abaxial view　G. Tepals of staminate flower, adaxial view　H. Tepals of staminate flower, abaxial view　I. Pistillate flower, adaxial view　J. Pistillate flower, lateral view　K. Tepals of pistillate flower, adaxial view　L. Tepals of pistillate flower, abaxial view　M. Androecium, adaxial view　N. Capsule　O. Ovary, transversely sectioned　P. Gynoecium, adaxial view　Q. Styles and stigmas

红探戈秋海棠

Begonia 'Red Tango'

　　该品种由荷兰育种者安东·胡夫纳格尔斯（Antoon Hoefnagels）培育，具体时间不详。

　　该品种叶色犹如魔法般多变，同一植株上通常可以看到暗绿色和红色、紫色、银色的混合配色，观赏价值高，适应性强，在华南地区室外露天和办公室内均可栽培。

Dutch breeder Antoon Hoefnagels created this cultivar. It exhibits a magical and ever-changing leaf coloration, combining dark green, red, purple, and silver hues often seen on the same plant. It holds high ornamental value and adapts well to various environments, making it suitable for outdoor cultivation in the south China and indoor cultivation in office spaces.

A

3cm

A. 植株　B. 花序　C. 叶片正反面　D. 雄花正面观　E. 雄花反面观　F. 雄花花被片正面观　G. 雄花花被片反面观　H. 雌花正面观　I. 雌花反面观　J. 雌花花被片正面观　K. 雌花花被片反面观　L. 雄蕊　M. 蒴果　N. 子房横切面　O. 雌蕊　P. 花柱和柱头

A. Habit　B. Inflorescence　C. Leaf, adaxial and abaxial view　D. Staminate flower, adaxial view　E. Staminate flower, abaxial view　F. Tepals of staminate flower, adaxial view G. Tepals of staminate flower, abaxial view　H. Pistillate flower, adaxial view　I. Pistillate flower, abaxial view　J. Tepals of pistillate flower, adaxial view　K. Tepals of pistillate flower, abaxial view　L. Androecium, adaxial view　M. Capsule　N. Ovary, transversely sectioned O. Gynoecium, lateral view　P. Styles and stigmas

海蓝宝石秋海棠

Begonia 'Aquamarine'

　　该品种由楚格（Zug）于1953年培育，母本是路德维格秋海棠（*Begonia ludwigii*），父本是西尔瓦多秋海棠（*Begonia* 'Silvadore'）。

　　该品种株型紧凑，叶片正面褐绿色被银白色斑纹，背面褐红色。

The cultivar, bred by Zug in 1953, crosses *Begonia ludwigii* as the female parent and *Begonia* 'Silvadore' as the male parent. It has a compact growth habit, with brown-green leaves on the adaxial surface with silver-white markings, and reddish-brown on the abaxial surface.

A

10cm

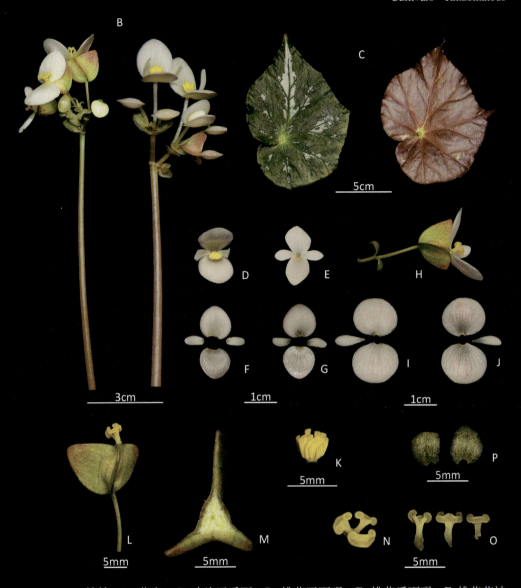

A. 植株　B. 花序　C. 叶片正反面　D. 雄花正面观　E. 雄花反面观　F. 雄花花被片正面观　G. 雄花花被片反面观　H. 雌花侧面观　I. 雌花花被片正面观　J. 雌花花被片反面观　K. 雄蕊　L. 蒴果　M.子房横切面　N. 雌蕊　O. 花柱和柱头　P.苞片

A. Habit　B. Inflorescence　C. Leaf, adaxial and abaxial view　D. Staminate flower, adaxial view　E. Staminate flower, abaxial view　F. Tepals of staminate flower, adaxial view　G. Tepals of staminate flower, abaxial view　H. Pistillate flower, lateral view　I. Tepals of pistillate flower, adaxial view　J. Tepals of pistillate flower, abaxial view　K. Androecium, lateral view　L. Capsule　M. Ovary, transversely sectioned　N. Gynoecium, adaxial view　O. Styles and stigmas　P. Bracts

秋之华秋海棠

Begonia 'Autumn Ember'

　　该品种由美国Logee Greenhouses公司于2014年从马默杜克秋海棠（*Begonia* 'Marmaduke'）和天使之光秋海棠（*Begonia* 'Angel Glow'）杂交种子苗选育出来的品种。

　　该品种叶片橙红色，花序高出植株，花量大，花被片淡粉色，在阴凉处生长良好，但也能忍受阳光照射。

The cultivar was developed by Logee Greenhouses in the United States in 2014 through the hybridization of *Begonia* 'Marmaduke' and *Begonia* 'Angel Glow'. Its leaves are orange-red, producing large flower clusters with pale pink tepals. It thrives in shady conditions but can also tolerate some sunlight.

A

3cm

A. 植株　B. 花序　C. 小苞片　D. 叶片正反面　E. 叶片横切面　F. 雄花正面观 G. 雄花反面观　H. 雄花花被片正面观　I. 雄花花被片反面观　J. 雌花正面观　K. 雌花 反面观　L. 雌花花被片正面观　M. 雌花花被片反面观　N. 雄蕊　O. 蒴果　P. 子房横 切面　Q. 雌蕊　R. 花柱和柱头

A. Habit　B. Inflorescence　C. Bracteole　D. Leaf, adaxial and abaxial view　E. Leaf, transversely sectioned　F. Staminate flower, adaxial view　G. Staminate flower, abaxial view H. Tepals of staminate flower, adaxial view　I. Tepals of staminate flower, abaxial view J. Pistillate flower, adaxial view　K. Pistillate flower, abaxial view　L. Tepals of pistillate flower, adaxial view　M. Tepals of pistillate flower, abaxial view　N. Androecium, lateral view　O. Capsule P. Ovary, transversely sectioned　Q. Gynoecium, adaxial view　R. Styles and stigmas

伯利恒之星秋海棠

Begonia 'Bethlehem Star'

该品种由美国育种者华莱士·瓦格纳（Wallace W. Wagner）于1975年培育，母本为睫毛秋海棠（*Begonia bowerae*），父本为艾普利秋海棠（*Begonia* 'Eppley'）。

该品种叶片呈墨绿色，叶缘被白色睫毛状长毛，耐热、耐湿。

The cultivar was developed by American breeder Wallace W. Wagner in 1975, with *Begonia bowerae* as the female parent and *Begonia* 'Eppley' as the male parent. The leaves of this cultivar are dark green with white long eyelash-like hairs along the margins. It is heat-tolerant and can withstand humid conditions.

A

2cm

A. 植株　B. 花序　C. 叶片正反面　D. 雄花正面观　E. 雄花反面观　F. 雄花花被片正面观　G. 雄花花被片反面观　H. 雌花正面观　I. 雌花侧面观　J. 雌花花被片正面观　K. 雌花花被片反面观　L. 雄蕊　M 蒴果　N. 子房横切面　O. 雌蕊　P. 花柱和柱头　Q. 苞片

A. Habit　B. Inflorescence　C. Leaf, adaxial and abaxial view　D. Staminate flower, adaxial view　E. Staminate flower, abaxial view　F. Tepals of staminate flower, adaxial view G. Tepals of staminate flower, abaxial view　H. Pistillate flower, adaxial view　I. Pistillate flower, lateral view　J. Tepals of pistillate flower, adaxial view K. Tepals of pistillate flower, abaxial view　L. Androecium, lateral view　M. Capsules　N. Ovary, transversely sectioned O. Gynoecium, lateral view　P. Styles and stigmas　Q. Bracts, adaxial and abaxial view

男友秋海棠

Begonia 'Boy Friend'

　　该品种由日本育种者御园勇（Isamu Misono）于1978年育成，母本为贝叶秋海棠（*Begonia conchifolia*），父本为雅致小姐秋海棠（*Begonia* 'Dainty Lady'）。

　　该品种叶片盾状，薄肉质，表面绿色，叶缘墨绿色，叶部有一玫红色小点，花序高于植株，花被片玫粉色。

The cultivar was developed by Japanese breeder Isamu Misono in 1978, with *Begonia conchifolia* as the female parent and *Begonia* 'Dainty Lady' as the male parent. Its leaves are shield-shaped, thin and fleshy, with a green surface and dark green leaf margins, and feature small crimson dots. The inflorescences rise above the plant, and tepals are pink.

A

10cm

A. 植株　B. 花序　C. 苞片　D. 叶片正反面　E. 雄花侧面观　F. 雄花花被片正面观　G. 雄花花被片反面观　H. 雌花正面观　I. 雌花侧面观　J. 雌花花被片正面观　K. 雌花花被片反面观　L. 雄蕊　M. 蒴果　N. 子房横切面　O. 雌蕊　P. 花柱和柱头

A. Habit　B. Inflorescence　C. Bracts　D. Leaf, adaxial and abaxial view, Petiole
E. Staminate flower, lateral view　F. Tepals of staminate flower, adaxial view　G. Tepals of staminate flower, abaxial view　H. Pistillate flower, adaxial view　I. Pistillate flower, lateral view　J. Tepals of pistillate flower, adaxial view　K. Tepals of pistillate flower, abaxial view L. Androecium, lateral view　M. Capsule　N. Ovary, transversely sectioned　O. Gynoecium, adaxial view　P. Styles and stigmas

克娄巴特拉秋海棠

Begonia 'Cleopatra'

　　该品种由美国育种者沃克（Walker）于1960年培育，母本为马菲尔秋海棠（*Begonia* 'Maphil'），父本为黑美人秋海棠（*Begonia* 'Black Beauty'）。

　　该品种叶片掌状中裂，似枫叶，表面草绿色，叶脉周围和叶缘点缀褐绿色斑纹，花被片淡粉色。

The cultivar was produced by American breeder Walker in 1960, with *Begonia* 'Maphil' as the female parent and *Begonia* 'Black Beauty' as the male parent. The leaves are palmately lobed, resembling maple leaves, with a grass-green surface. The leaf veins and margins are adorned with brown-green markings, and tepals are pale pink.

A

5cm

A. 植株　B. 花序　C.叶片正反面　D. 雄花正面观　E. 雄花侧面观　F. 雄花花被片正面观　G. 雄花花被片反面观　H. 雌花正面观　I. 雌花侧面观　J. 雌花花被片正面观　K. 雌花花被片反面观　L. 雄蕊　M. 蒴果　N.子房横切面　O. 雌蕊　P. 花柱和柱头

A. Habit　B. Inflorescence　C. Leaf, adaxial and abaxial view　D. Staminate flower, adaxial view　E. Staminate flower, lateral view　F. Tepals of staminate flower, adaxial view　G. Tepals of staminate flower, abaxial view　H. Pistillate flower, adaxial view　I. Pistillate flower, lateral view　J. Tepals of pistillate flower, adaxial view　K. Tepals of pistillate flower, abaxial view　L. Androecium, lateral view　M. Capsule　N. Ovary, transversely sectioned　O. Gynoecium, adaxial view　P. Styles and stigmas

黛西秋海棠
Begonia 'Daisy'

该品种由美国育种者保罗·洛（Paul P. Lowe）于1972年育成，母本为马氏秋海棠（*Begonia* 'Chumash'），父本为睫毛秋海棠（*Begonia bowerae*）。

该品种抗性强，耐热，叶片古铜色，具有天鹅绒质地，叶缘具白色睫毛状长毛。

The cultivar was developed by American breeder Paul P. Lowe in 1972, with *Begonia* 'Chumash' as the female parent and *Begonia bowerae* as the male parent. It exhibits strong resistance and heat tolerance. Its leaves have an antique copper color and a velvety texture, with white eyelash-like margins.

A

5cm

A. 植株　B. 花序　C. 苞片　D. 叶片正反面，叶柄　E. 雄花正面观　F. 雄花反面观　G. 雄花花被片正面观　H. 雄花花被片反面观　I. 雌花正面观　J. 雌花侧面观　K. 雌花花被片正面观　L. 雌花花被片反面观　M. 雄蕊　N. 蒴果　O. 子房横切面　P. 雌蕊　Q. 花柱和柱头

A. Habit　B. Inflorescence　C. Bracts　D. Leaf, adaxial and abaxial view, petiole　E. Staminate flower, adaxial view　F. Staminate flower, abaxial view　G. Tepals of staminate flower, adaxial view　H. Tepals of staminate flower, abaxial view　I. Pistillate flower, adaxial view　J. Pistillate flower, lateral view　K. Tepals of pistillate flower, adaxial view　L. Tepals of pistillate flower, abaxial view　M. Androecium, lateral view　N. Capsule　O. Ovary, transversely sectioned　P. Gynoecium, adaxial view　Q. Styles and stigmas

红叶秋海棠

Begonia 'Erythrophylla'

该品种别名牛排秋海棠，由波兰育种者约瑟夫·瓦尔塞维兹·拉维茨（Joseph Warsceiwitz Ritter von Rawicz）于1845年培育，母本为天胡荽叶秋海棠（*Begonia hydrocotylifolia*），父本为长袖秋海棠（*Begonia manicata*），是一个最古老的且至今依然在全世界广泛栽培的品种。

该品种叶片光滑呈椭圆形，正面为橄榄绿色，背面为酒红色，叶片上有明显的星形脉络。

The cultivar, known as 'Beefsteak Begonia', was developed by Polish breeder Joseph Warsceiwitz Ritter von Rawicz in 1845. Its parent plants were *Begonia hydrocotylifolia* and *Begonia manicata*. This is one of the oldest and most widely cultivated varieties worldwide. The leaves of this cultivar are smooth and oval-shaped, with a glossy olive-green color on the adaxial surface and a wine-red color on the abaxial surface. The leaves exhibit prominent star-shaped venation.

A

10cm

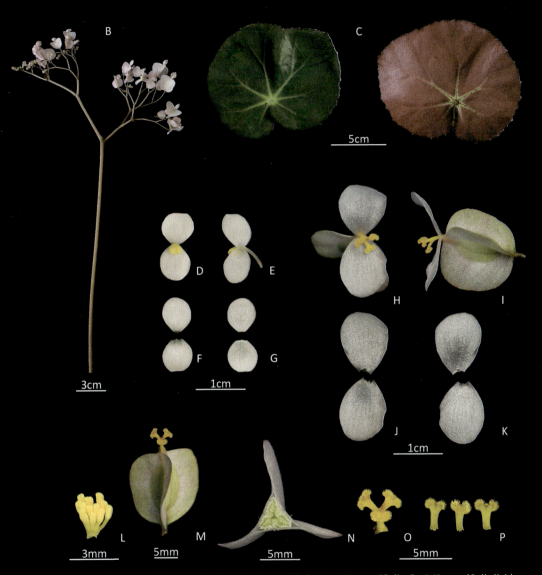

A. 植株　B. 花序　C. 叶片正反面　D. 雄花正面观　E. 雄花反面观　F. 雄花花被片正面观　G. 雄花花被片反面观　H. 雌花正面观　I. 雌花侧面观　J. 雌花花被片正面观　K. 雌花花被片反面观　L. 雄蕊　M. 蒴果　N. 子房横切面　O. 雌蕊　P. 花柱和柱头

A. Habit　B. Inflorescence　C. Leaf, adaxial and abaxial view　D. Staminate flower, adaxial view　E. Staminate flower, abaxial view　F. Tepals of staminate flower, adaxial view G. Tepals of staminate flower, abaxial view　H. Pistillate flower, adaxial view　I. Pistillate flower, lateral view　J. Tepals of pistillate flower, adaxial view　K. Tepals of pistillate flower, abaxial view　L. Androecium, lateral view　M. Capsule　N. Ovary, transversely sectioned O. Gynoecium, lateral view　P. Styles and stigmas

象形秋海棠

Begonia 'Heiroglyphics'

　　该品种由美国育种者布拉德·汤普森（Brad Thompson）于2001年育成，母本为克娄巴特拉秋海棠（*Begonia* 'Cleopatra'），父本为 [*Begonia* '（'Bokit' × *carrieae*)'] 。

　　该品种叶片卵圆形，掌状中裂，叶基微螺旋状，叶缘具白色睫毛状长毛。表面绿色，沿叶缘具棕红色斑纹。花梗挺出植株，花被片淡粉色。

The cultivar was developed by American breeder Brad Thompson in 2001, with *Begonia* 'Cleopatra' as the female parent and *Begonia* '('Bokit' × *carrieae*)' as the male parent. Its leaves are ovate, palmately lobed, and have a slightly spiraled base. The leaf margins are adorned with white-colored hairs, and the adaxial surface is green, with reddish-brown markings along the margins.

A

5cm

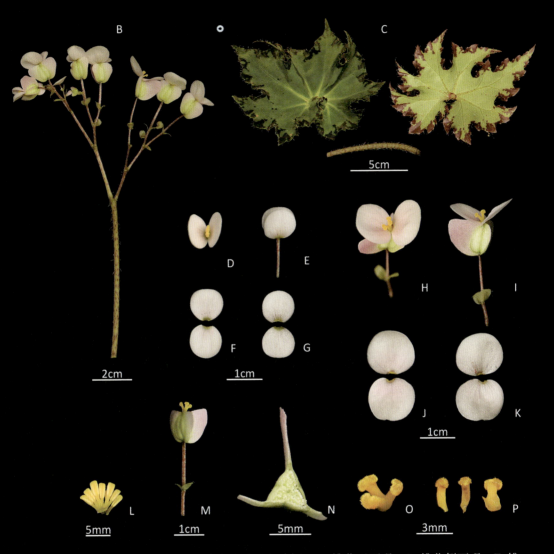

A. 植株　B. 花序　C. 叶片正反面和叶柄　D. 雄花正面观　E. 雄花侧面观　F. 雄花花被片正面观　G. 雄花花被片反面观　H. 雌花正面观　I. 雌花侧面观　J. 雌花花被片正面观　K. 雌花花被片反面观　L. 雄蕊　M. 蒴果　N. 子房横切面　O. 雌蕊　P. 花柱和柱头

A. Habit　B. Inflorescence　C. Leaf, adaxial and abaxial view, and petiole　D. Staminate flower, adaxial view　E. Staminate flower, lateral view　F. Tepals of staminate flower, adaxial view　G. Tepals of staminate flower, abaxial view　H. Pistillate flower, adaxial view　I. Pistillate flower, lateral view　J. Tepals of pistillate flower, adaxial view　K. Tepals of pistillate flower, abaxial view　L. Androecium, lateral view　M. Capsule　N. Ovary, transversely sectioned　O. Gynoecium, adaxial view　P. Styles and stigmas

273

牛仔舞秋海棠

Begonia '**Jive**'

该品种由美国育种者于2016年育成，亲本未知。

该品种叶片卵圆形，掌状浅裂，叶缘具白色睫毛状长毛，表面草绿色，沿叶缘具棕红色斑纹。

The cultivar was developed by an American breeder in 2016, and the parentage is unknown. Its leaves are ovate, palmately lobed, and have white-colored hairs along the margins. The leaf surface is grass green, with reddish-brown markings along the margins.

A

5cm

A. 植株　B. 花序　C. 叶片正反面　D. 雄花正面观　E. 雄花反面观　F. 雄花花被片正面观　G. 雄花花被片反面观　H. 雌花正面观　I. 雌花侧面观　J. 雌花花被片正面观　K. 雌花花被片反面观　L. 雄蕊　M. 蒴果　N. 子房横切面　O. 雌蕊　P. 花柱和柱头

A. Habit　B. Inflorescence　C. Leaf, adaxial and abaxial view　D. Staminate flower, adaxial view　E. Staminate flower, adaxial view　F. Tepals of staminate flower, adaxial view G. Tepals of staminate flower, abaxial view　H. Pistillate flower, adaxial view　I. Pistillate flower, lateral view　J. Tepals of pistillate flower, adaxial view　K. Tepals of pistillate flower, abaxial view　L. Androecium, lateral view　M. Capsule　N. Ovary, transversely sectioned O. Gynoecium, adaxial view　P. Styles and stigmas

小兄弟秋海棠

Begonia 'Little Brother Montgomery'

该品种由美国育种者马丁·约翰逊（Martin Johnson）于1979年育成。母本为裂叶秋海棠（*Begonia palmata*），父本为王冠秋海棠（*Begonia diadema*）。

该品种具有地上直立茎，叶片掌状浅裂，叶片银白色，叶缘和叶基紫褐色，背面紫红色。

The cultivar, developed by American breeder Martin Johnson in 1979, has *Begonia palmata* as the female parent and *Begonia diadema* as the male parent. It features upright jointed stems and palmately lobed leaves with a silvery-white color. The leaf margins and base are purplish-brown, and the abaxial surface of the leaves is purplish-red.

A

5cm

A. 植株　B. 花序　C. 叶片正反面　D. 雄花侧面观　E. 雄花花被片正面观　F. 雄花花被片反面观　G. 雌花正面观　H. 雌花反面观　I. 雌花花被片正面观　J. 雌花花被片反面观　K. 雄蕊　L. 蒴果　M. 子房横切面　N. 雌蕊　O. 花柱和柱头

A. Habit　B. Inflorescence　C. Leaf, adaxial and abaxial view　D. Staminate flower, lateral view　E. Tepals of staminate flower, adaxial view　F. Tepals of staminate flower, abaxial view　G. Pistillate flower, adaxial view　H. Pistillate flower, abaxial view　I. Tepals of pistillate flower, adaxial view　J. Tepals of pistillate flower, abaxial view　K. Androecium, lateral view　L. Capsule　M. Ovary, transversely sectioned　N. Gynoecium, adaxial view O. Styles and stigmas

幻影秋海棠

Begonia 'Mirage'

该品种中文别名玛雅阁秋海棠，由美国育种者帕特里克·沃利（Patrick J. Worley）于1984年育成，亲本未知。

该品种叶片表面灰绿色，背面红色，子房玫红色的翅尤其显眼，花量大，耐热性极强。

The cultivar was developed by American breeder Patrick J. Worley in 1984, with the parentage unknown. The leaves of this cultivar have a grayish-green surface and are red on the back, with particularly striking pink wings on the ovary. It produces abundant flowers and exhibits excellent heat tolerance.

A

10cm

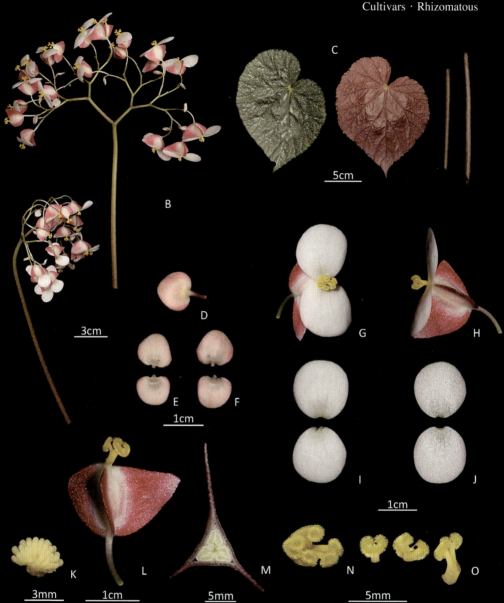

A. 植株　B. 花序　C. 叶片正反面和叶柄　D. 雄花侧面观　E. 雄花花被片正面观
F. 雄花花被片反面观　G. 雌花正面观　H. 雌花侧面观　I. 雌花花被片正面观　J. 雌花花
被片反面观　K. 雄蕊　L. 蒴果　M. 子房横切面　N. 雌蕊　O. 花柱和柱头

A. Habit　B. Inflorescence　C. Leaf, adaxial and abaxial view, and petiole　D. Staminate
flower, lateral view　E. Tepals of staminate flower, adaxial view　F. Tepals of staminate flower, abaxial
view　G. Pistillate flower, adaxial view　H. Pistillate flower, lateral view　I. Tepals of pistillate flower,
adaxial view　J. Tepals of pistillate flower, abaxial view　K. Androecium, lateral view　L. Capsules
M. Ovary, transversely sectioned　N. Gynoecium, adaxial view　O. Styles and stigmas

滑雪座秋海棠

Begonia 'New Skeezar'

该品种于1984年培育而成，母本为*Begonia ludicra*，父本为褐脉秋海棠（*Begonia glandulosa*）。

该品种叶片表面光滑，密被白色，如白雪覆盖叶面。

The cultivar was developed in 1984, with *Begonia ludicra* as the female parent and *Begonia glandulosa* as the male parent. It has smooth leaves with a dense white covering, resembling a snow-covered leaf surface.

A

5cm

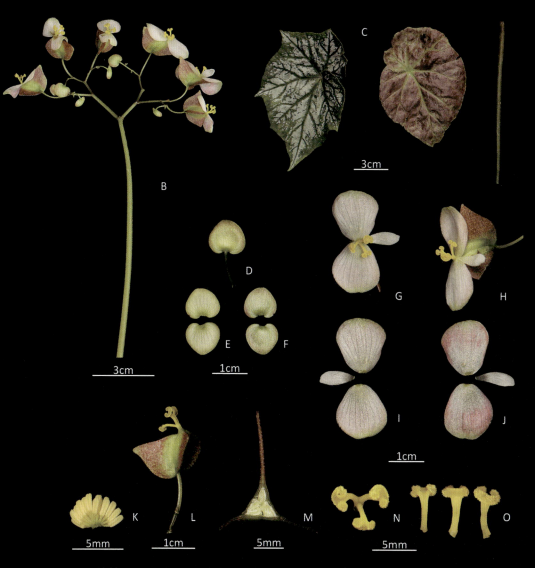

A. 植株　B. 花序　C. 叶片正反面和叶柄　D. 雄花侧面观　E. 雄花花被片正面观 F. 雄花花被片反面观　G. 雌花正面观　H. 雌花反面观　I. 雌花花被片正面观　J. 雌花 花被片反面观　K. 雄蕊　L. 蒴果　M. 子房横切面　N. 雌蕊　O. 花柱和柱头

A. Habit　B. Inflorescence　C. Leaf, adaxial and abaxial view, and petiole D. Staminate flower, lateral view　E. Tepals of staminate flower, adaxial view　F. Tepals of staminate flower, abaxial view　G. Pistillate flower, adaxial view　H. Pistillate flower, abaxial view　I. Tepals of pistillate flower, adaxial view　J. Tepals of pistillate flower, abaxial view K. Androecium, lateral view　L. Capsule　M. Ovary, transversely sectioned　N. Gynoecium, adaxial view　O. Styles and stigmas

宁明银秋海棠

Begonia ningmingensis 'Ningming Silver'

该品种由上海辰山植物园田代科等于2017年从宁明秋海棠（*Begonia ningmingensis*）中优选出的银色个体。

该品种株型好，全株被毛，叶片密被白斑而成银色，花期长，适应性良好。

The cultivar was selected from *Begonia ningmingensis* by Tian Daike and others at Shanghai Chenshan Botanical Garden in 2017. It has a well-formed habit and is covered by hairs. The leaves are densely adorned with white spots, giving them a silvery appearance. It has a long flowering period and exhibits excellent adaptability.

A

10cm

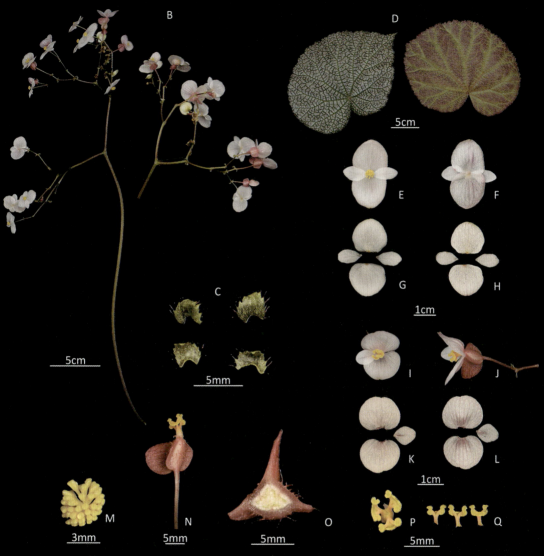

A. 植株　B. 花序　C. 苞片　D. 叶片正反面　E. 雄花正面观　F. 雄花反面观
G. 雄花花被片正面观　H. 雄花花被片反面观　I. 雌花正面观　J. 雌花侧面观　K. 雌花
花被片正面观　L. 雌花花被片反面观　M. 雄蕊　N. 蒴果　O. 子房横切面　P. 雌蕊
Q. 花柱和柱头

A. Habit　B. Inflorescence　C. Bracts　D. Leaf, adaxial and abaxial view　E. Staminate
flower, adaxial view　F. Staminate flower, abaxial view　G. Tepals of staminate flower,
adaxial view　H. Tepals of staminate flower, abaxial view　I. Pistillate flower, adaxial
view　J. Pistillate flower, lateral view　K. Tepals of pistillate flower, adaxial view　L. Tepals
of pistillate flower, abaxial view　M. Androecium, adaxial view　N. Capsule　O. Ovary,
transversely sectioned　P. Gynoecium, adaxial view　Q. Styles and stigmas

晚秋霜秋海棠

Begonia 'November Frost'

该品种由美国育种者莱斯利·伍德里夫（Leslie Woodriff）于1982年育成，母本为帝王秋海棠（*Begonia imperialis*），父本为博克特秋海棠（*Begonia* 'Bokit'）。

该品种叶片螺旋状，叶片正面密被白色斑点，背面酒红色。

The cultivar was developed by American breeder Leslie Woodriff in 1982, with *Begonia imperialis* as the female parent and *Begonia* 'Bokit' as the male parent. The leaves are spirally arranged, with the adaxial surface covered in white spots, while the abaxial surface is a rich maroon color.

A

5cm

A. 植株　B. 花序　C. 苞片　D. 叶片正反面　E. 雄花花被片正面观　F. 雄花花被片反面观　G. 雌花正面观　H. 雌花反面观　I. 雌花花被片正面观　J. 雌花花被片反面观　K. 雄蕊　L. 蒴果　M. 子房横切面　N. 雌蕊　O. 花柱和柱头

A. Habit　B. Inflorescence　C. Bracts, adaxial and abaxial view　D. Leaf, adaxial and abaxial view　E. Tepals of staminate flower, adaxial view　F. Tepals of staminate flower, abaxial view　G. Pistillate flower, adaxial view　H. Pistillate flower, abaxial view　I. Tepals of pistillate flower, adaxial view　J. Tepals of pistillate flower, abaxial view　K. Androecium, lateral view　L. Capsule　M. Ovary, transversely sectioned　N. Gynoecium, adaxial view　O. Styles and stigmas

帕拉玛王子秋海棠

Begonia 'Palomar Prince'

该品种由美国育种者迈克尔·卡图兹（Michael Kartuz）于1998年育成，母本为瓦尔蒙秋海棠（*Begonia* 'Valmont'），父本未知。

该品种深绿色的叶子呈双螺旋形，叶面点缀着浅绿色的斑点，亮色的叶脉粗而明显。

The cultivar was developed by American breeder Michael Kartuz in 1998, with *Begonia* 'Valmont' as the female parent and an unknown male parent. It features deep green leaves with a double spiral shape. The leaf surface is adorned with light green spots, and the bright-colored veins are thick and prominent.

A

10cm

A. 植株　B. 花序　C. 苞片　D. 叶片正反面和叶柄　E. 雄花正面观　F. 雄花花被片正面观　G. 雄花花被片反面观　H. 雌花正面观　I. 雌花侧面观　J. 雌花花被片正面观　K. 雌花花被片反面观　L. 雄蕊　M. 蒴果　N. 子房横切面　O. 雌蕊　P. 花柱和柱头

A. Habit　B. Inflorescence　C. Bracts, adaxial and abaxial view　D. Leaf, adaxial and abaxial view, and petiole　E. Staminate flower, adaxial view　F. Tepals of staminate flower, adaxial view　G. Tepals of staminate flower, abaxial view　H. Pistillate flower, adaxial view　I. Pistillate flower, lateral view　J. Tepals of pistillate flower, adaxial view　K. Tepals of pistillate flower, abaxial view　L. Androecium, lateral view　M. Capsule　N. Ovary, transversely sectioned　O. Gynoecium, adaxial view　P. Styles and stigmas

瞬息风暴秋海棠

Begonia 'Passing Storm'

该品种由美国育种者帕特里克·沃利（Patrick J. Worley）于1983年育成，并于2001年在美国秋海棠协会登录。

该品种具有叶脉呈灰色，叶片呈现出粉红色的光泽，随着叶子的生长，颜色会从粉红色变为银色。

The cultivar was developed by American breeder Patrick J. Worley in 1983 and was registered with the American Begonia Society in 2001. It has grayish-veined leaves with a pink sheen, and as the leaves mature, their color changes from pink to silver.

A

5cm

A. 植株　B. 花序　C. 叶片正反面　D. 雄花侧面观　E. 雄花花被片正面观　F. 雄花花被片反面观　G. 雌花正面观　H. 雌花侧面观　I. 雌花花被片正面观　J. 雌花花被片反面观　K. 雄蕊　L. 蒴果　M. 子房横切面　N. 雌蕊　O. 花柱和柱头

A. Habit　B. Inflorescence　C. Leaf, adaxial and abaxial view　D. Staminate flower, lateral view　E. Tepals of staminate flower, adaxial view　F. Tepals of staminate flower, abaxial view　G. Pistillate flower, adaxial view　H. Pistillate flower, lateral view　I. Tepals of pistillate flower, adaxial view　J. Tepals of pistillate flower, abaxial view　K. Androecium, lateral view　L. Capsule　M. Ovary, transversely sectioned　N. Gynoecium, adaxial view O. Styles and stigmas

银色御园秋海棠

Begonia 'Silver Misono'

　　该品种由日本育种者御园勇（Isamu Misono）于1978年育成，母本为银宝石秋海棠（*Begonia* 'Silver Jewell'），父本爱丽丝秋海棠（*Begonia alice-clarkiae*）。

　　该品种叶子呈亮绿色，沿着叶脉分布有银色斑点，质感如毛毯。

The cultivar was hybridized by Japanese breeder Isamu Misono in 1978, using *Begonia* 'Silver Jewell' as the female parent and *Begonia alice-clarkiae* as the male parent. It has bright green leaves with silver splashes distributed along the veins, giving it a velvety texture akin to a blanket.

A

10cm

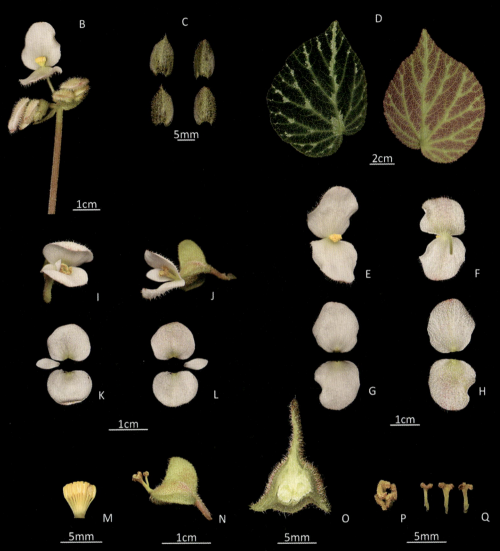

A. 植株　B. 花序　C. 苞片　D. 叶片正反面　E. 雄花正面观　F. 雄花反面观
G. 雄花花被片正面观　H. 雄花花被片反面观　I. 雌花正面观　J. 雌花侧面观　K. 雌花
花被片正面观　L. 雌花花被片反面观　M. 雄蕊　N. 蒴果　O. 子房横切面　P. 雌蕊
Q. 花柱和柱头

A. Habit　B. Inflorescence　C. Bracts, adaxial and abaxial view　D. Leaf, adaxial and abaxial view　E. Staminate flower, adaxial view　F. Staminate flower, abaxial view　G. Tepals of staminate flower, adaxial view　H. Tepals of staminate flower, abaxial view　I. Pistillate flower, adaxial view　J. Pistillate flower, lateral view　K. Tepals of pistillate flower, adaxial view　L. Tepals of pistillate flower, abaxial view　M. Androecium, lateral view　N. Capsule O. Ovary, transversely sectioned　P. Gynoecium, adaxial view　Q. Styles and stigmas

斯波尔丁秋海棠

Begonia 'Spaulding'

　　该品种异名为利莲安秋海棠（*Begonia* 'Lillian Steinhouse'），由美国育种者路易丝·施韦特费格（Louise Schwerdtfeger）于1951年育成，母本为睫毛秋海棠（*Begonia bowerae*），父本是天胡荽叶秋海棠（*Begonia hydrocotylifolia*）。

　　该品种具有匍匐的根状茎，叶片小巧，边缘密被睫毛状长纤毛，靠近叶片边缘的叶脉呈现出黑褐色。

The cultivar synonym is *Begonia* 'Lillian Steinhouse'. It was hybridized by American breeder Louise Schwerdtfeger in 1951, with *Begonia bowerae* as the female plant and *Begonia hydrocotylifolia* as the male plant. It features creeping rhizomes and small, attractive leaves with densely arranged long cilia-like hairs along the margin. The leaf veins near the leaf margin exhibit a dark brown coloration.

A

5cm

A. 植株　B. 花序　C. 苞片　D. 叶片正反面　E. 雄花正面观　F. 雄花反面观
G. 雄花花被片正面观　H. 雄花花被片反面观　I. 雌花侧面观　J. 雌花花被片正面观
K. 雌花花被片反面观　L. 雄蕊　M. 蒴果　N. 子房横切面　O. 雌蕊　P. 花柱和柱头

A. Habit　B. Inflorescence　C. Bracts, adaxial and abaxial view　D. Leaf, adaxial and
abaxial view　E. Staminate flower, adaxial view　F. Staminate flower, abaxial view　G. Tepals
of staminate flower, adaxial view　H. Tepals of staminate flower, abaxial view　I. Pistillate
flower, lateral view　J. Tepals of pistillate flower, adaxial view　K. Tepals of pistillate flower,
abaxial view　L. Androecium, lateral view　M. Capsule　N. Ovary, transversely sectioned
O. Gynoecium, adaxial view　P. Styles and stigmas

镜叶秋海棠

Begonia 'Speculata'

　　该品种叶片呈暗灰绿色，表面有皱褶效果；叶脉呈亮绿色，沿叶脉两侧分布较多白色小斑点；花白色，子房呈绿色，绿色小苞片宿存。育种信息不详。

The cultivar's leaves are dark gray-green color with a wrinkled surface; the leaf veins are bright green and have numerous small white spots along both sides of the veins. The flowers are white, and the ovaries are green, with persistent small green bracts. Breeding information is not detailed.

A

10cm

A. 植株　B. 花序　C. 叶片正反面　D. 雄花正面观　E. 雄花侧面观　F. 雄花花被片正面观　G. 雄花花被片反面观　H. 雌花正面观　I. 雌花侧面观　J. 雌花花被片正面观　K. 雌花花被片反面观　L. 雄蕊　M. 蒴果　N. 子房横切面　O. 雌蕊　P. 花柱和柱头

A. Habit　B. Inflorescence　C. Leaf, adaxial and abaxial view　D. Staminate flower, adaxial view　E. Staminate flower, lateral view　F. Tepals of staminate flower, adaxial view　G. Tepals of staminate flower, abaxial view　H. Pistillate flower, adaxial view　I. Pistillate flower, lateral view　J. Tepals of pistillate flower, adaxial view　K. Tepals of pistillate flower, abaxial view　L. Androecium, lateral view　M. Capsule　N. Ovary, transversely sectioned　O. Gynoecium, adaxial view　P. Styles and stigmas

虎斑秋海棠

Begonia 'Tiger Kitten'

该品种由美国育种者莱斯利·伍德里夫（Leslie Woodriff）于1973年育成，父母本信息不详。

该品种植株小巧，具有许多小圆形的叶片，呈现出绿色和深紫红色的鲜明花纹。

属于睫毛秋海棠的品种。

The cultivar was hybridized by American breeder Leslie Woodriff in 1973, and the parentage information is unknown. It is small and compact, with numerous small, ovate leaves that exhibit vibrant green and deep burgundy patterns. It belongs to the group of _Begonia bowerae_.

A

5cm

A. 植株　B. 花序　C. 叶片正反面和叶柄　D. 雄花正面观　E. 雄花反面观　F. 雄花花被片正面观　G. 雄花花被片反面观　H. 雌花正面观　I. 雌花侧面观　J. 雌花花被片正面观　K. 雌花花被片反面观　L. 雄蕊　M. 蒴果　N. 子房横切面　O. 雌蕊　P. 花柱和柱头

A. Habit　B. Inflorescence　C. Leaf, adaxial and abaxial view, and petiole　D. Staminate flower, adaxial view　E. Staminate flower, abaxial view　F. Tepals of staminate flower, adaxial view　G. Tepals of staminate flower, abaxial view　H. Pistillate flower, adaxial view　I. Pistillate flower, lateral view　J. Tepals of pistillate flower, adaxial view　K. Tepals of pistillate flower, abaxial view　L. Androecium, lateral view　M. Capsule　N. Ovary, transversely sectioned　O. Gynoecium, adaxial view　P. Styles and stigmas

U400秋海棠

Begonia U400

　　属于未鉴定的秋海棠系列种类之一，据美国秋海棠协会官方网站介绍，该种（或品种）最早由亨利·拉波特（Henri Laporte）从马来西亚的Songkhla市场购买，并被带到美国栽培种植。

　　U400秋海棠叶片椭圆形，除叶脉和叶缘呈现灰绿色外，叶片正面全为银色。该种（或品种）耐热、耐湿，耐晒，环境适应性极好。

　　It is one of the unidentified species. According to the American Begonia Society (ABS), this species (or cultivar) was initially obtained by Henri Laporte from the Songkhla Market in Malaysia and brought to the United States for cultivation. The leaves of *Begonia* U400 are oval-shaped, with a silver color on the entire adaxial surface, except for the leaf veins and margins, which appear grayish-green. This species (or cultivar) exhibits high heat and humidity tolerance and can withstand direct sunlight, making it highly adaptable to various environmental conditions.

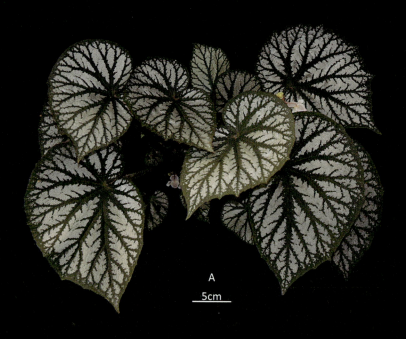

A

5cm

A. 植株　B. 花序　C. 叶片正反面和叶柄　D. 雄花正面观　E. 雄花反面观　F. 雄花花被片正面观　G. 雄花花被片反面观　H. 雄蕊

A. Habit　B. Inflorescence　C. Leaf, adaxial and abaxial view, and petiole　D. Staminate flower, adaxial view　E. Staminate flower, abaxial view　F. Tepals of staminate flower, adaxial view　G. Tepals of staminate flower, abaxial view　H. Androecium, adaxial view

红毛柄秋海棠

Begonia 'Verschaffeltii'

该品种由德国育种者爱德华·奥古斯特·冯·雷格尔（Eduard August von Regel）于1853年育成，母本为环毛秋海棠（*Begonia manicata*），父本为瓜栗叶秋海棠（*Begonia carolineifolia*）。

该品种叶片卵形，叶缘尖锐分裂，呈浅橄榄绿色，叶柄、叶脉覆盖着短而坚硬的红粉色毛。二歧伞状花序，高于叶片，花呈淡玫瑰色。

The German breeder Eduard August von Regel developed the cultivar in 1853, with *Begonia manicata* as the female parent and *Begonia carolineifolia* as the male parent. The leaves are ovate, with sharply denticulate margins, and exhibit light olive-green color. The petioles and veins of the leaves are covered with short, stiff, reddish-pink hairs. The inflorescences are dichotomous cyme, elevated above the leaves, and the flowers are pale rose.

A

10cm

A. 植株　B. 花序　C. 叶片正反面　D. 雄花正面观　E. 雄花花被片正面观　F. 雄花花被片反面观　G. 雌花侧面观　H. 雌花正面观　I. 雌花花被片正面观　J. 雌花花被片反面观　K. 雄蕊　L. 蒴果　M. 子房横切面　N. 雌蕊　O. 花柱和柱头

A. Habit　B. Inflorescence　C. Leaf, adaxial and abaxial view　D. Staminate flower, adaxial view　E. Tepals of staminate flower, adaxial view　F. Tepals of staminate flower, abaxial view　G. Pistillate flower　lateral view　H. Pistillate flower, adaxial view　I. Tepals of pistillate flower, adaxial view　J. Tepals of pistillate flower, abaxial view　K. Androecium, lateral view　L. Capsule　M. Ovary, transversely sectioned　N. Gynoecium, adaxial view　O. Styles and stigmas

闪电秋海棠

Begonia 'Lightning'

　　该品种由澳大利亚育种者罗斯·鲍尔韦尔（Ross Bolwell）于1983年以奥尔森秋海棠（*Begonia olsoniae*）为父本与一个未知的母本杂交得到。

　　该品种叶片长卵形，表面深绿色，背面酒红色，具有非常明显的亮绿色叶脉。

The cultivar was hybridized by Australian breeder Ross Bolwell in 1983, crossing *Begonia olsoniae* with an unknown female parent. It has oblong-ovate leaves, with a deep green color on the adaxial surface and a wine-red color on the abaxial surface, featuring distinct bright green veins.

A

10cm

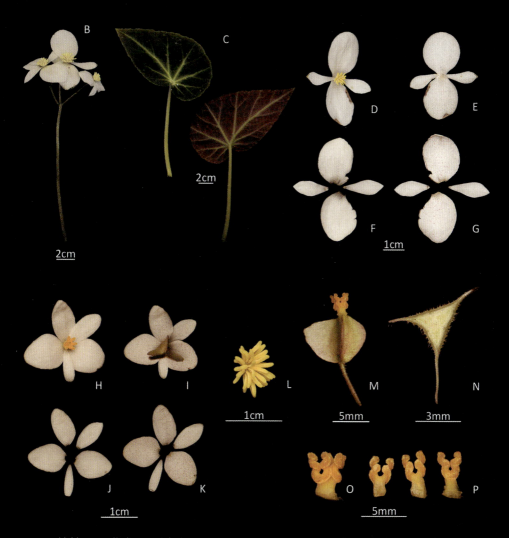

A. 植株　B. 花序　C.叶片正反面　D. 雄花正面观　E. 雄花反面观　F. 雄花花被片正面观　G. 雄花花被片反面观　H. 雌花正面观　I. 雌花反面观　J. 雌花花被片正面观　K. 雌花花被片反面观　L. 雄蕊　M. 蒴果　N. 子房横切面　O. 雌蕊　P. 花柱和柱头

A. Habit　B. Inflorescence　C. Leaf, adaxial and abaxial view　D. Staminate flower, adaxial view　E. Staminate flower, abaxial view　F. Tepals of staminate flower, adaxial view　G. Tepals of staminate flower, abaxial view　H. Pistillate flower, adaxial view　I. Pistillate flower, abaxial view　J. Tepals of pistillate flower, adaxial view　K. Tepals of pistillate flower, abaxial view　L. Androecium, adaxial view　M. Capsule　N. Ovary, transversely sectioned　O. Gynoecium, lateral view　P. Styles and stigmas

博尼塔秋海棠

Begonia 'Bonita Shea'

  该品种与玛瑙斯（*Begonia* 'Manaus'）及 *Begonia* U002疑为同一种，为自然杂交种，母本为红芒秋海棠（*Begonia thelmae*），父本为巴西变色秋海棠（*Begonia solimutata*）。

  该品种具有匍匐茎，叶片较小，卵圆形，正反面均被糙毛，黄绿色叶脉明显。

  The cultivar is suspected to be the same as *Begonia* 'Manaus' and *Begonia* U002, representing a natural hybrid. The female parent is *Begonia thelmae*, while the male parent is *Begonia solimutata*, both native to Brazil. It exhibits creeping rhizomes with relatively small, oblong-ovate leaves, covered with scabrous hairs on both surfaces. The leaf veins are prominently yellow-green.

A

A. 植株　B. 花序　C. 叶片正反面　D. 雄花正面观　E. 雄花反面观　F. 雄花花被片正面观　G. 雄花花被片反面观　H. 雌花正面观　I. 雌花反面观　J. 雌花花被片正面观　K. 雌花花被片反面观　L. 雄蕊　M. 蒴果　N. 子房横切面　O. 雌蕊　P. 花柱和柱头

A. Habit　B. Inflorescence　C. Leaf, adaxial and abaxial view　D. Staminate flower, adaxial view　E. Staminate flower, abaxial view　F. Tepals of staminate flower, adaxial view　G. Tepals of staminate flower, abaxial view　H. Pistillate flower, adaxial view　I. Pistillate flower, abaxial view　J. Tepals of pistillate flower, adaxial view　K. Tepals of pistillate flower, abaxial view　L. Androecium, lateral view　M. Capsule　N. Ovary, transversely sectioned　O. Gynoecium, adaxial view　P. Styles and stigmas

中文检索表
Chinese Name Index

拉丁名检索表
Scientific Name Index

图书在版编目（CIP）数据

秋海棠属植物形态解剖图鉴／李凌飞，杨蕾蕾主编
．—北京：中国农业出版社，2023.12（2024.12重印）
ISBN 978-7-109-31617-1

Ⅰ．①秋… Ⅱ．①李… ②杨… Ⅲ．①秋海棠科－植
物解剖学－图集 Ⅳ.①Q949.72-64

中国国家版本馆CIP数据核字（2024）第015433号

秋海棠属植物形态解剖图鉴
QIUHAITANG SHU ZHIWU XINGTAI JIEPOU TU-JIAN

中国农业出版社出版
地址：北京市朝阳区麦子店街18号楼
邮编：100125
责任编辑：丁瑞华
版式设计：王　晨　　责任校对：吴丽婷　　责任印制：王　宏
印刷：北京中科印刷有限公司
版次：2023年12月第1版
印次：2024年12月北京第2次印刷
发行：新华书店北京发行所
开本：700mm×1000mm　1/16
印张：20.25
字数：385千字
定价：198.00元